Elements
of
Structural Dynamics

Elements
of
Structural Dynamics

Glen V. Berg

Professor Emeritus of Civil Engineering
The University of Michigan, Ann Arbor

Prentice Hall
Englewood Cliffs, New Jersey 07632

Library of Congress Cataloging-in-Publication Data

BERG, GLEN V.
    Elements of structural dynamics/Glen V. Berg.
        p. cm. — (Prentice Hall international series in civil
    engineering and engineering mechanics)
    Includes index.
    ISBN 0-13-272493-6
    1. Structural dynamics. I. Title. II. Series.
TA654.B47 1989
624.1'71 — dc19   88-6981
                                                       CIP

The author and publisher of this book have used their best efforts in preparing this book. These efforts include the development, research, and testing of the theories and programs to determine their effectiveness. The author and publisher make no warranty of any kind, expressed or implied, with regard to these programs or the documentation contained in this book. The author and publisher shall not be liable in any event for incidental or consequential damages in connection with, or arising out of, the furnishing, performance, or use of these programs.

Editorial/production supervision and
    interior design: Debbie Young
Cover design: Ben Santora
Manufacturing buyer: Mary Noonan

© 1989 by Prentice-Hall, Inc.
A Division of Simon & Schuster
Englewood Cliffs, New Jersey 07632

Printed in the United States of America

10 9 8 7 6 5 4 3 2 1

ISBN 0-13-272493-6

Prentice-Hall International (UK) Limited, *London*
Prentice-Hall of Australia Pty. Limited, *Sydney*
Prentice-Hall Canada Inc., *Toronto*
Prentice-Hall Hispanoamericana, S.A., *Mexico*
Prentice-Hall of India Private Limited, *New Delhi*
Prentice-Hall of Japan, Inc., *Tokyo*
Simon & Schuster Asia Pte. Ltd., *Singapore*
Editora Prentice-Hall do Brasil, Ltda., *Rio de Janeiro*

To my students

Contents

Preface

This book stems from a course in structural dynamics taught to first-year graduate students at The University of Michigan for more than two decades. While the basic principles have changed little over that period, computational techniques have changed tremendously and much new knowledge has been gained, especially new knowledge of earthquakes and their effects on structures. Earthquake and wind are the two most common dynamic loads on buildings and civil-engineering structures. The text uses the earthquake problem as the main vehicle for conveying the principles and procedures to be explored.

The text presumes that the reader knows the elements of structural analysis and design through the usual undergraduate curriculum, including statically indeterminate analysis, as well as mathematics through elementary differential equations and elementary matrix algebra. Some familiarity with computer programming is helpful but not absolutely essential. While most of the examples can be solved with an electronic hand calculator and a sufficient quantity of patience and perseverance, a computer is most helpful. BASIC programs for most of the numerical procedures are presented, since it is the most common language of the personal computer today. Conversion to other languages, such as FORTRAN, Pascal, or C, can be readily accomplished.

The level of mathematics is kept as elementary as the subject matter will reasonably allow. Emphasis throughout is on basic principles and practical procedures. Accordingly, some derivations are less concise or less elegant than they might be made with the use of more sophisticated mathematical techniques. We hope they are also less obscure.

Elements
of
Structural Dynamics

CHAPTER ONE

Basic Concepts

1.1 NEWTON'S SECOND LAW

Sir Isaac Newton (1642–1727) postulated the laws of motion that form the basis of structural dynamics. Newton spoke of the "quantity of motion" of a body, the product of its mass times its velocity, which we now designate as its momentum. He postulated that if a force acts upon a body, the rate of change of the quantity of motion of the body is equal to the applied force. That is,

$$d(mv)/dt = f \qquad (1.1.1)$$

in which both the momentum mv and the driving force f are functions of time. This is Newton's Second Law, the cornerstone of structural dynamics.

Newton also formulated a First Law, that a body in motion with no forces acting on it will continue to move in a straight line and at a constant velocity, and a Third Law, that for every action there is an equal and opposite reaction.

1

In structural dynamics, we deal with systems in which the mass m is constant. For such systems,

$$f = m\,dv/dt = ma \tag{1.1.2}$$

where the driving force f, the velocity v, and the acceleration a are functions of time. Equation (1.1.2) is the familiar "force equals mass times acceleration" equation. For it to be valid, the mass must be constant. Systems for which the mass varies with time are not necessarily esoteric or complex. An example would be a rocket or other vehicle for which some of the mass (the fuel) is consumed by the process that generates the driving force. We will arbitrarily exclude such systems from our consideration. Newton's Second Law also carries the restriction that it is valid only for particle velocities that are small compared with the velocity of light. That restriction will not bother us.

In this text, we will use a Newtonian notation, denoting time derivatives by dots over a variable, or by Roman superscripts for time derivatives higher than the third. Thus,

$$\dot{x} = dx/dt \qquad \ddot{x} = d^2x/dt^2 \qquad \dddot{x} = d^3x/dt^3 \qquad x^{iv} = d^4x/dt^4 \qquad \text{etc.}$$

In this notation, Eq. (1.1.2) becomes

$$f = m\ddot{x} \tag{1.1.3}$$

which applies to all particles of mass in a dynamic system. If the physical system is represented mathematically as a continuum, there are infinitely many mass particles, and, therefore, an infinite number of coordinates would be needed to define the position of all the mass in the system. The system would have infinitely many *degrees of freedom*.

1.2 DEGREES OF FREEDOM

The number of **degrees of freedom** in a dynamic system is the least number of coordinates needed to define the position of all of the particles of mass in the system.

In most cases, we can justifiably discard some components of motion as being insignificant compared with others, and thereby get a simplified dynamic system with fewer degrees of freedom. For example, consider the pendulum shown in Fig. 1.1, consisting of a rock suspended from a fixed point by a bar. If we took the bar and the rock to be deformable and to have continuous distributions of mass, then the system would have infinitely many degrees of freedom. However, if the bar were much lighter than the rock, we might neglect its mass altogether, and if the displacements of the

Figure 1.1 Pendulum.

particles of the rock due to its change in shape were likely to be small compared with those due to its movement as a body, we might consider the rock to be infinitely rigid. These assumptions would reduce the number of degrees of freedom to six, which in Cartesian coordinates would be three translations and three rotations with respect to the coordinate axes. If the size of the rock were small compared with the length of the bar, we might treat the rock as a point mass. That would reduce the number of degrees of freedom to three, the three components of displacement of the mass. Finally, if we took the bar to be inextensible and took all motion to be in one plane, we would remove two more degrees of freedom, leaving only one, the rotation of the bar about the hinge.

A comparable case would be the spring–mass system of Fig. 1.2. If the mass of the spring were distributed along its length, or if the suspended mass were deformable, then there would be infinitely many degrees of freedom. But if we considered the spring to be weightless, the mass to be a point mass, and all movement to be in the vertical direction, then we would have once again a single-degree-of-freedom dynamic system.

A more complex case is the four-story plane frame shown in Fig. 1.3(a). Again the system with its distributed mass would have infinitely many degrees of freedom. However, if we idealized the frame as a weightless structural system, with the structural components retaining all of their stiffness properties but having no weight, replaced all the mass of the system by four rigid lumped masses attached to the weightless frame at the four floor levels, as in Fig. 1.3(b), and then took all motion of the lumped masses to be

Figure 1.2 Spring–mass system.

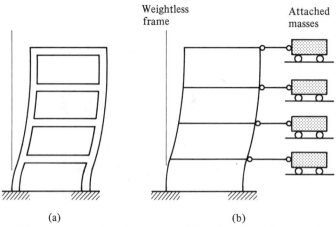

Weightless frame   Attached masses

(a)   (b)

Figure 1.3 Four-story plane frame. (a) Flexible frame with distributed mass. (b) Idealization.

in the horizontal direction in the plane of the frame, then we would have replaced the original system by an "equivalent" system having four degrees of freedom, the horizontal displacements of the four masses. This kind of idealization will be useful indeed.

In most, but not all, of the cases we shall consider, it will be possible to idealize the system as a number of discrete lumped masses attached to a weightless structure. To write the equations of motion, we shall consider each mass to be a free body acted upon by external driving forces and internal restoring and damping forces, and derive the equations of motion for each mass from a consideration of Newton's Second Law.

1.3 MASS AND WEIGHT

The difference between **mass** and **weight** is sometimes confusing, less so in Standard International (SI) units than in either centimeter–gram–second metric (cgs) units or English units.

Mass is a measure of the quantity of matter. **Weight** is a measure of the force necessary to impart a specified acceleration to a specified mass. The **acceleration of gravity,** denoted **g,** is the acceleration that the gravity of the earth would impart to a free-falling body at sea level.

The English system is based on fundamental units of force, length, and time, that is, the **pound, foot** (or inch), and **second**, respectively. Other units are derived from these. The usual unit of mass is the **pound mass** (**lbm**), that mass to which the **pound force** (**lbf** or **lb**) will impart an acceleration equal to the acceleration of gravity (1 g). The acceleration of gravity is 32.174 ft/sec^2 or 386.1 in./sec^2. Thus,

$$1 \text{ lbm} = \frac{1 \text{ lbf}}{32.174 \text{ ft/sec}^2}$$

One pound mass *weighs* 1 pound force. If you place the pound mass on the scale, the needle will register 1 pound force. This is quite different from saying that 1 pound mass is equal to 1 pound force. They are not equal, for the pound mass has units of $FL^{-1}T^2$ and the pound force has units of F.

A unit of mass sometimes used in the English system, especially in the field of hydraulics, is the **slug**, defined as that mass to which a force of 1 pound would impart an acceleration of 1 ft/sec^2. Thus,

$$1 \text{ slug} = \frac{1 \text{ lbf}}{1 \text{ ft/sec}^2} = \frac{32.174 \text{ lbf}}{1 \text{ g}} = 32.174 \text{ lbm}$$

This has the merit of consistency. A unit force imparts a unit acceleration to a unit mass. Alas, the consistency does not permeate the English system, and the consistency fails to carry through to other measures, such as stress. Structural dynamics rarely sees the slug used. One slug is *equal* to 32.174 lbm. One slug *weighs* 32.174 lbf.

The cgs metric system employs fundamental units of mass, length, and time, that is, the **gram** (or **kilogram**), **centimeter** (or **meter**), and **second**, respectively. The kilogram and meter are used more often in structural dynamics than the gram and centimeter. Other units are derived from these. The usual unit of force is the **kilogram force (kgf)**, which is the force that would impart to 1 **kilogram mass** (**kgm** or **kg**) an acceleration equal to the acceleration of gravity (1 g). The acceleration of gravity is 9.807 m/sec^2. Thus, in cgs metric units,

$$1 \text{ kgf} = 1 \text{ kg} \cdot 9.807 \text{ m/sec}^2$$

A mass of 1 kilogram weighs 1 kilogram force. Again, a kilogram mass is not equal to a kilogram force, for they are measured in different units, the kilogram mass in units of M and the kilogram force in units of MLT^{-2}.

The Standard International (SI) system of units is a consistent system, and some day, if it is eventually adopted, it may do away with some of the confusion attendant on the English and cgs metric systems. The SI system employs for its fundamental units of mass, length, and time, the **kilogram (kg)**, **meter (m)**, and **second (sec)**, respectively. Other units are derived from these. The unit of force is the **Newton (N)**, defined as that force that would impart to a mass of 1 kg an acceleration of 1 m/sec^2. Thus,

$$1 \text{ N} = 1 \text{ kg} \cdot 1 \text{ m/sec}^2$$

Note the consistency. A unit force imparts a unit acceleration to a unit mass. Of course, we have this much consistency with the English system if we use the slug as the unit of mass. However, in SI units, the consistency carries through to other measures as well.

In dynamics even more than in statics, it is helpful to carry units along with the mathematical operations. Relations to remember are

$$\text{Force} = \text{mass} \times \text{length} \times \text{time}^{-2}$$
$$\text{Mass} = \text{force} \times \text{length}^{-1} \times \text{time}^{2}$$

1.4 THE IMPULSE–MOMENTUM RELATION

The impulse–momentum relation is an extension of Newton's Second Law, or an alternate form of it. Newton's Second Law is

$$f = m\ddot{x} = m\, d^2 x / dt^2 \tag{1.1.3}$$

Integrate both sides with respect to t to get

$$\int_{t_1}^{t_2} f\, dt = m(\dot{x}_2 - \dot{x}_1) \tag{1.4.1}$$

The integral on the left side of Eq. (1.4.1) is the area under the force–time curve, which is the impulse. The product of mass and velocity is momentum. Thus, we get the relation

$$\text{Impulse} = \text{change in momentum}$$

There is nothing particularly profound about this, and it accomplishes nothing that cannot be accomplished with the equation of motion. For some problems, the impulse–momentum relation may provide a useful alternative approach.

For example, consider the following problem, shown in Fig. 1.4. A rigid block of weight W is initially at rest on a horizontal plane. The coefficient of Coulomb friction between the block and the plane is μ, independent of velocity. An initial peak triangular force pulse of initial value P_0 and duration t_1 acts on the block. Find (a) the time t_2 at which the block comes to rest, and (b) the distance x_2 that the block moves.

One method of solution would be to derive and solve the differential equation of motion. If we consider the block to be a free body with driving force f and friction force r acting on it, we can derive the equation of motion from Newton's law, obtaining

$$f - r = (W/g)\ddot{x} \tag{1.4.2}$$

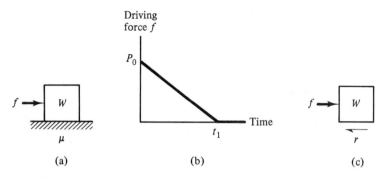

Figure 1.4   Block driven by triangular force pulse. (a) System. (b) Driving force. (c) Free-body diagram.

in which the driving force

$$f = P_0(1 - t/t_1) \qquad 0 < t < t_1$$
$$f = 0 \qquad t \geq t_1$$

The friction force is somewhat more elusive. There are three possibilities: (a) if the block is stationary and the driving force is inadequate to overcome friction, then the friction force exactly balances the driving force; (b) if the block is stationary and the driving force is great enough to overcome friction, then the friction force is the coefficient of friction times the weight of the block; and (c) if the block is in motion, then the friction force is the coefficient of friction times the weight. The block moves only in the positive direction, that is, in the direction of the driving force. Expressed mathematically, these possible states are

$$\text{if } \dot{x} = 0 \quad \text{and} \quad f < \mu W, \qquad \text{then } r = f$$
$$\text{if } \dot{x} > 0 \quad \text{or} \quad f \geq \mu W, \qquad \text{then } r = \mu W$$

and

$$\dot{x} \geq 0 \qquad \text{always}$$

The initial conditions are $x = 0$ and $\dot{x} = 0$.

To solve Eq. (1.4.2) with these constraints and initial conditions is not especially difficult, but the several possibilities make the process tedious. The force may or may not be great enough to overcome friction. If it does overcome friction, then the block will move, but it may finally come to a stop either before or after the force pulse ends.

The impulse–momentum relation provides a simple alternative. Combining the triangular driving force f with the friction force r, we find the net force $f - r$ to be one of the three cases shown in Fig. 1.5.

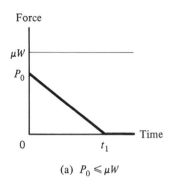

(a) $P_0 \leqslant \mu W$

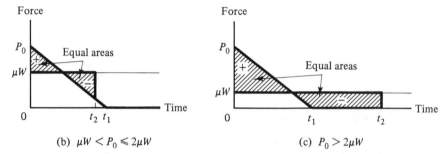

(b) $\mu W < P_0 \leqslant 2\mu W$ (c) $P_0 > 2\mu W$

Figure 1.5 Three possible cases for example of Fig. 1.4. (a) No motion. (b) Motion stops before pulse ends. (c) Motion continues after pulse ends.

In the case of Fig. 1.5(a), the initial driving force is inadequate to overcome friction, and there is no motion at all; hence, $t_2 = 0$ and $x_2 = 0$.

If the initial driving force overcomes friction, motion continues until the velocity returns to zero. The initial momentum was zero; the momentum when motion ceases is zero. Thus, the change in momentum is zero, so the area under the net force-vs.-time curve must be zero. In the case of Fig. 1.5(b), motion ceases before the end of the pulse, and from geometry, we get $t_2 = 2(1 - \mu w/P_0)t_1$.

In the case of Fig. 1.5(c), the pulse ends before motion ceases, and again geometry gives us the terminal time, which is $t_2 = (P_0/2\mu w)t_1$.

For the remaining question, how far does the block move, the acceleration of the block is the net force divided by the mass. The structural engineer may find it useful to recognize that for a system initially at rest, integrating acceleration once to get velocity and a second time to get displacement is procedurally the same as integrating load once to get shear and a second time to get bending moment. The zero initial velocity and displacement of the dynamic problem correspond to the zero shear and bending moment at the free end of a cantilever beam. Hence, if we take the static moment of the area under the acceleration curve up to any time t, about t, we get the displacement at time t. Thus, for the three cases shown in Fig. 1.5, we take the static moment about t_2 of the area under the net-force diagram to the left of t_2 and divide it by the mass W/g to get the displacement at time t_2:

Case (a): $P_0 \leq \mu W$ $x_2 = 0$ (1.4.3)

Case (b): $\mu W < P_0 \leq 2\mu W$ $x_2 = \dfrac{2(P_0 - \mu W)^3 g t_1^2}{3P_0^2 W}$ (1.4.4)

Case (c): $P_0 > 2\mu W$ $x_2 = \dfrac{P_0(3P_0 - 4\mu W)g t_1^2}{24\mu W^2}$ (1.4.5)

The impulse–momentum relation has enabled us to replace a differential-equation problem with a geometric problem.

1.5 THE WORK–ENERGY RELATION

Another useful formulation of the equation of motion can be obtained from work and energy considerations. Multiply both sides of Eq. (1.1.3) by $\dot{x}$ and integrate with respect to time to get

$$\int_{t_1}^{t_2} f\dot{x}\, dt = \int_{t_1}^{t_2} m\ddot{x}\dot{x}\, dt \qquad (1.5.1)$$

Because $\dot{x}\, dt = dx$ and $\ddot{x}\, dt = d\dot{x}$, this can be written

$$\int_{x_1}^{x_2} f\, dx = m\dot{x}_2^2/2 - m\dot{x}_1^2/2 \qquad (1.5.2)$$

The integral on the left side of Eq. (1.5.2) is the area under the force–displacement curve, the work done by the force f, and the two terms on the right side are the final and initial kinetic energy of the mass. Hence, we obtain the relation, perhaps intuitively obvious, **work = change in energy.** This work–energy relation, like the impulse–momentum relation, is no more powerful than the underlying equation of motion, but sometimes it is more convenient to use. Consider the example shown in Fig. 1.6. A rigid block of weight W is released from its state of rest positioned at a height h above a weightless spring of stiffness k, Fig. 1.6(a). Find the maximum deflection of the spring.

As before, the equation of motion can be derived by considering a free-body diagram of the block with all forces acting on it, Fig. 1.6(b), and then employing Newton's Second Law. The gravity force is a constant W. If

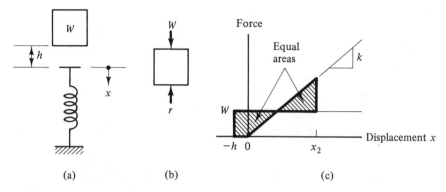

Figure 1.6 Rigid block dropped on weightless spring. (a) System. (b) Forces.
(c) Force–displacement relations.

we define x to be the position of the mass, measured downward from the
contact position, the spring force is

$$r = 0 \qquad x \leq 0$$
$$r = kx \qquad x > 0$$
$$(1.5.3)$$

The equation of motion is

$$W - r = (W/g)\ddot{x} \qquad (1.5.4)$$

and the initial conditions are $x_0 = -h$ and $\dot{x}_0 = 0$. We could solve
Eq. (1.5.4) in two stages, the first stage for the free fall and the second for
the time when the mass is in contact with the spring. The first solution
would give the initial conditions for the second. Then, by finding the time
of zero velocity in the second-stage solution, we could get the maximum
displacement. This would be straightforward, but not convenient.

The work–energy relation provides an attractive alternative. The net
downward force acting on the block as a function of displacement is the
weight less the spring force, as indicated in Fig. 1.6(b). Initially, the kinetic
energy was zero; at maximum displacement, it is again zero. For zero
change in energy, the work done must be zero; hence, the net area under the
force–displacement curve must be zero. The geometry of Fig. 1.6(b) gives
$W(h + x_2) - kx_2^2/2 = 0$. Defining δ to be the static displacement, $\delta = W/k$,
we can write the solution as

$$x_2 = \delta[1 + (1 + 2h/\delta)^{1/2}] \qquad (1.5.5)$$

1.6 SINGLE-DEGREE-OF-FREEDOM DYNAMIC SYSTEMS

Consider the classical spring–mass–damper system shown in Fig. 1.7. If we consider the spring and damper to be weightless, the mass to be rigid, and all motion to be in the direction of the x-axis, we have the classical linear damped single-degree-of-freedom (SDF) system. A single displacement co-ordinate describes the position of all the mass in the system.

For later convenience, we will let $u(t)$ be the displacement of the mass in the direction of the x-axis.

The spring is a linear spring of stiffness k, exerting a force ku, the **restoring force.** The damper is a viscous damper, shown schematically as a dashpot, and exerts a force $c\dot{u}$, the **damping force.** The **driving force** $f(t)$ is taken to be positive in the direction of the x-axis. Thus, if the driving force, the displacement, and the velocity are all positive, $f(t)$ acts in the positive direction of the x-axis and ku and $c\dot{u}$ act in the negative direction. Newton's Second Law then gives the equation of motion:

$$f(t) - ku - c\dot{u} = m\ddot{u} \qquad (1.6.1)$$

or

$$m\ddot{u} + c\dot{u} + ku = f(t) \qquad (1.6.2)$$

Equation (1.6.2) is the classical form of the equation of motion for the SDF damped linear dynamic system.

Gravity forces had no effect in the system of Fig. 1.7 because we took motion to be horizontal. To see the effect of gravity, let us incline the system, as shown in Fig. 1.8(a). Again, the mass is rigid and the spring weightless, all motion is in the direction of the spring, and there is no friction between the mass and the plane on which it moves. Denoting the displacement from the unstrained position as z, we get the forces on the

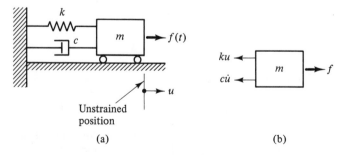

Figure 1.7   Spring–mass–damper system. (a) System. (b) Forces.

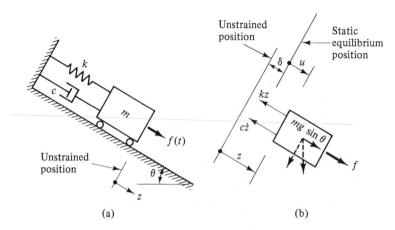

Figure 1.8 Inclined spring–mass–damper system. (a) System. (b) Forces.

displaced mass to be those of Fig. 1.8(b). Newton's Second Law gives us the equation of motion in the direction of the force to be

$$f(t) + mg \sin \theta - c\dot{z} - kz = m\ddot{z} \tag{1.6.3}$$

or

$$m\ddot{z} + c\dot{z} + kz = f(t) + mg \sin \theta \tag{1.6.4}$$

This time, the equation is complicated by the appearance of the force of gravity among its terms. However, we can eliminate the gravity term by measuring the displacement from the static-equilibrium position instead of the unstrained position. Let the static displacement of the mass under the force of gravity be δ, which is

$$\delta = (mg \sin \theta)/k \tag{1.6.5}$$

Then u, the displacement measured from the static equilibrium position, is

$$u = z - \delta$$

or

$$z = u + \delta$$

$$\dot{z} = \dot{u}$$

and

$$\ddot{z} = \ddot{u}$$

We substitute these in Eq. (1.6.4) to get

$$m\ddot{u} + c\dot{u} + ku + k\delta = f(t) + mg \sin \theta \tag{1.6.6}$$

But now the $k\delta$ term on the left side of Eq. (1.6.6) cancels the $mg \sin \theta$ term on the right, leaving the classical SDF equation of motion:

$$m\ddot{u} + c\dot{u} + ku = f(t) \tag{1.6.2}$$

Not only has the gravity term been removed, but also the equation is independent of the angle of inclination. Whether the plane is vertical, horizontal, or inclined, the equation of motion is the same as long as the displacement is measured from the position of static equilibrium.

Consider the pendulum shown in Fig. 1.9. A small mass m, considered to be a point mass, is suspended from a fixed point by a weightless inextensible bar of length L, and constrained to move only in the plane of the paper. A single coordinate θ, denoting the rotation of the bar from the vertical, serves to define the position of all the mass in the system; hence, it is a single-degree-of-freedom system.

In free vibration, the forces acting on the displaced mass are the tension in the bar and the vertical force of gravity. There is no radial motion, so the radial component of the bar tension offsets the radial component of the gravitational and inertia forces. The tangential component $mg \sin \theta$ accelerates the mass. Newton's Second Law gives us

$$-mg \sin \theta = mL\ddot{\theta}$$

or

$$\ddot{\theta} + (g/L) \sin \theta = 0 \tag{1.6.7}$$

This nonlinear differential equation describes the motion for any rotation θ, including complete revolutions. If we impose the restriction that displacements always remain small ($\theta \ll 1$), then $\sin \theta = \theta$, and Eq. (1.6.7) becomes

$$\ddot{\theta} + (g/L)\theta = 0 \tag{1.6.8}$$

which brings us again to the classical differential equation of motion for an undamped linear SDF system.

A mass on a weightless beam leads to an equivalent result. Consider the structure of Fig. 1.10, consisting of a rectangular platform supported on four identical columns oriented with their principal axes parallel to the edges of the platform, subjected to a driving force acting on the platform in one of the principal directions.

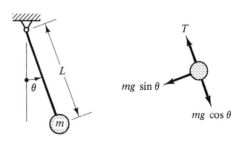

Figure 1.9   Ideal pendulum.
(a) System. (b) Forces.   (a)   (b)

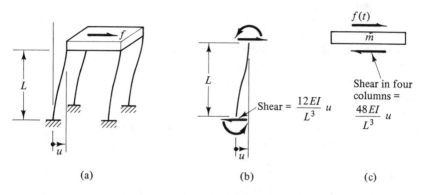

Figure 1.10   Rigid platform on four columns. (a) System. (b) Forces on displaced column. (c) Horizontal forces acting on mass.

If we consider the columns to be weightless and inextensible and the platform rigid, take motion to be only in the principal direction in which the driving force acts, neglect the effect of axial load on the stiffness of the columns, and neglect bending moments due to eccentricity of the gravity load, then we can derive the forces acting on the mass from a free-body diagram of a displaced column. The shear in each column, Fig. 1.10(b), is $12EI/L^3$ times the displacement. Four columns provide the total restoring force. From Newton's Second Law, we get

$$m\ddot{u} + (48EI/L^3)u = f(t) \qquad (1.6.9)$$

which is again the classical SDF equation.

1.7 DYNAMIC EQUILIBRIUM

Structural engineers, having been trained to think in terms of equilibrium of forces, may find d'Alembert's principle of dynamic equilibrium particularly useful. Jean le Rond d'Alembert (1717–1783), a French mathematician, developed the concept in 1742, when he introduced the notion of an inertia force, a force equal to the product of a mass times its acceleration and acting in a direction opposite to the acceleration. D'Alembert's principle states that with inertia forces included, a dynamic system is in equilibrium, that is, the sum of all of the forces (or moments) acting on the system, including the inertia forces (or moments), is zero. Figure 1.11 shows a free-body diagram of the displaced mass of Fig. 1.7, including the inertia force. Setting the sum of all of the forces equal to zero gives the equation of motion.

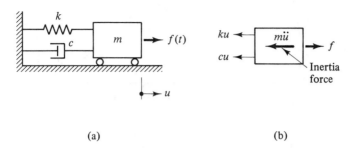

(a)   (b)

Figure 1.11 Dynamic equilibrium. (a) Spring–mass–damper system.
(b) Forces in dynamic equilibrium.

1.8 VIRTUAL DISPLACEMENTS

The principle of virtual displacements applies to dynamic as well as static
systems. According to this principle, if a system in dynamic equilibrium un-
dergoes an arbitrary small virtual displacement, the work done by all of the
forces (including the inertia force) is zero. This says nothing really new — it
is just a restatement of Newton's Second Law or d'Alembert's principle, but
in a form that is sometimes easier to apply. Thus, for the spring–mass–
damper system of Fig. 1.11, the work done by the driving force f undergo-
ing a virtual displacement δu is $f\,\delta u$; the work done by the spring force ku is
$-ku\,\delta u$; the work done by the damper force $c\dot{u}$ is $-c\dot{u}\,\delta u$; and the work done
by the inertia force $m\ddot{u}$ is $-m\ddot{u}\,\delta u$. Setting the total work to zero gives

$$(f - m\ddot{u} - c\dot{u} - ku)\,\delta u = 0 \qquad (1.8.1)$$

or

$$m\ddot{u} + c\dot{u} + ku = f \qquad (1.6.2)$$

1.9 MULTIPLE-MASS OR DISTRIBUTED-MASS SYSTEMS

The foregoing examples have all involved either single-point masses or the
translation of rigid distributed masses, which may be considered as a mass
lumped at a single point. This may not always be valid. Consider, for example, the system shown in Fig. 1.12(a), consisting of a rigid weightless bar
with two lumped masses attached to it, supported by a fulcrum, with an attached spring and damper, and acted upon by a driving force as shown. We
will consider only small displacements. This is a single-degree-of-freedom
system, but it is difficult to replace the two masses by an equivalent mass
lumped at a single point.

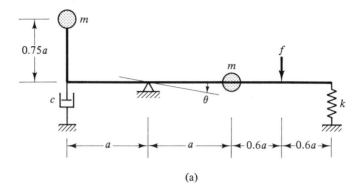

(a)

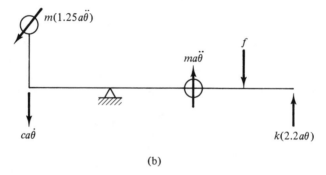

(b)

Figure 1.12 Two-mass single-degree-of-freedom system. (a) System. (b) Forces in dynamic equilibrium.

We have some latitude in choosing the displacement coordinate. Quite arbitrarily, we choose it to be θ, the clockwise rotation about the fulcrum. Figure 1.12(b) shows the forces acting on the system, including the inertia forces.

(a) *Dynamic equilibrium* requires that the sum of the moments about the fulcrum be zero. Thus,

$$-m(1.25a\ddot{\theta})(1.25a) - c(a\dot{\theta})(a) - m(a\ddot{\theta})(a)$$
$$+ f(1.6a) - k(2.2a\theta)(2.2a) = 0 \qquad (1.9.1)$$

or

$$2.5625ma\ddot{\theta} + ca\dot{\theta} + 4.84ka\theta = 1.6f(t) \qquad (1.9.2)$$

(b) The principle of *virtual displacements* may also be used. The work done by all of the forces of Fig. 1.12(b) undergoing a small virtual displacement $\delta\theta$ is

$$\delta W = k(2.2a\theta)(-2.2a\delta\theta) + f(1.6a\delta\theta) + m(a\ddot{\theta})(-a\delta\theta)$$
$$+ c(a\dot{\theta})(-a\delta\theta) + m(1.25a\ddot{\theta})(-1.25a\delta\theta) = 0 \tag{1.9.3}$$

With the common factor $a\delta\theta$ canceled, Eq. (1.9.3) becomes the same as Eq. (1.9.2).

(c) We may also use *work–energy relations* to derive the equation of motion. In the displaced configuration of the system, the energy in the conservative components of the system is the sum of the potential energy of the spring plus the kinetic energy of the two masses. These are

$$\text{Potential energy} = k(2.2a\theta)^2/2$$
$$\text{Kinetic energy} = m(a\dot{\theta})^2/2 + m(1.25a\dot{\theta})^2/2$$

In a small increment of time δt, the changes would be

$$\delta\text{PE} = d\text{PE}/dt\,\delta t = k(2.2a\theta)(2.2a)\dot{\theta}\,\delta t \tag{1.9.4}$$

$$\delta\text{KE} = d\text{KE}/dt\,\delta t = m(a\dot{\theta})(a)\ddot{\theta}\,\delta t + m(1.25a\dot{\theta})(1.25a)\ddot{\theta}\,\delta t \tag{1.9.5}$$

In this increment of time, the driving force f would displace a distance $1.6a\dot{\theta}\,\delta t$ in the direction of the force and the damper force $ca\dot{\theta}$ would displace a distance $a\dot{\theta}\,\delta t$ in a direction opposite to the force. Thus, the energy added to the system by these forces would be

$$f(1.6a\dot{\theta}\,\delta t) - c(a\dot{\theta})(a\dot{\theta}\,\delta t)$$

The net energy added to the system by the driving and damper forces must be equal to the change in potential and kinetic energy. Thus,

$$k(2.2a)^2\theta\dot{\theta}\,\delta t + m(2.5625a^2)\ddot{\theta}\dot{\theta}\,\delta t = f(1.6a)\dot{\theta}\,\delta t - ca^2\dot{\theta}^2\,\delta t \tag{1.9.6}$$

We delete the common factor $a\dot{\theta}\,\delta t$ from the terms of Eq. (1.9.6), thereby obtaining Eq. (1.9.2) again.

We could also derive Eq. (1.9.2) from Newton's Second Law, to be sure, but the derivation would be more laborious.

Consider a second example, the system of Fig. 1.13(a). Here we have a rigid square plate of mass per unit area m_0 hinged to a rigid bar of mass per unit length m_1, with a vertical spring at the hinge, a vertical damper at the midpoint of the bar, supported on two fulcrums, and acted upon by a uniformly distributed dynamic driving force acting horizontally on the vertical edge of the plate. Although the mass is distributed, there is only one degree of freedom. We choose the displacement coordinate to be θ, the clockwise rotation of the plate about the left support. Figure 1.13(b) shows the forces acting on the system, including the inertia forces. Because the plate and bar both translate as well as rotate, we must include both translational and rotational inertia forces for them. There are other ways of looking at the problem, but this is as convenient as any.

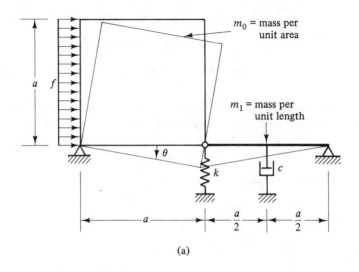

(a)

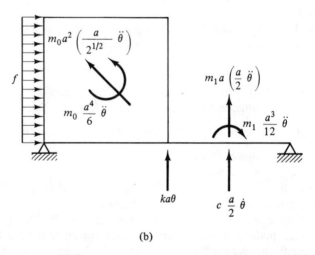

(b)

Figure 1.13 Single-degree-of-freedom system with distributed mass.
(a) System. (b) Dynamic equilibrium forces.

We use the principle of virtual displacements to derive the equation of motion. We impose a small virtual displacement $\delta\theta$ and set the work done by all of the forces equal to zero:

$$fa(a\,\delta\theta/2) - m_0a^2(a\ddot{\theta}/2^{1/2})(a\,\delta\theta/2^{1/2}) - m_0a^4\ddot{\theta}\,\delta\theta/6 - ka\theta a\,\delta\theta$$
$$- c(a\dot{\theta}/2)(a\,\delta\theta/2) - m_1a(a\ddot{\theta}/2)(a\,\delta\theta/2) \qquad (1.9.7)$$
$$- m_1(a^3/12)\ddot{\theta}\,\delta\theta = 0$$

which, after a common $a^2 \delta\theta$ is factored out, becomes

$$(2m_0 a^2/3 + m_1 a/3)\ddot{\theta} + (c/4)\dot{\theta} + k\theta = f/2 \qquad (1.9.8)$$

1.10 SYSTEMS WITH FLEXIBLE MASS

All the foregoing systems had the entire mass as either point masses or rigid bodies. If the system contains elements that have both mass and flexibility, generally it has infinitely many degrees of freedom. Later, we shall consider the analysis of such systems by the mathematically rigorous method of normal modes, but for the present, let us consider a means of approximating such a system as a linear SDF system.

If we can reasonably postulate that the flexible elements of a dynamic system deform in some fixed *shape,* and that only the *amplitude* of deformation varies with time, then we can describe the motion of the system by a single variable, and we have only one degree of freedom. In a sense, we have done this in previous examples, such as those of Figs. 1.12 and 1.13, where we took parts of the system to be infinitely rigid. Here, however, we shall assume deformation patterns other than rigid-body motions.

For example, suppose the spring–mass–damper system of Fig. 1.14 has a rigid mass and weightless damper, but the spring of stiffness k has a

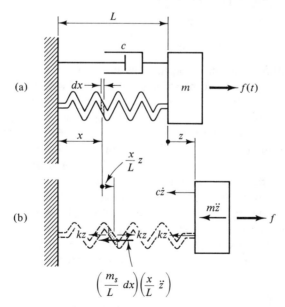

Figure 1.14 Spring–mass–damper system with distributed mass in spring. (a) System. (b) Forces.

mass m_s distributed uniformly along its length as well. There are then infinitely many degrees of freedom. However, let us assume that the spring is always uniformly stretched, that is, the tension in the spring at any instant, and therefore the strain, is uniform. We then have only one degree of freedom. To get the equation of motion, we invoke the principle of virtual displacements.

Figure 1.14 shows the system in both its unstrained and deformed positions. The displacement of the end mass is z and the displacement of the element of spring is $(x/L)z$. When the virtual displacement of the end mass is δz, the corresponding virtual displacement of the element of spring must be $(x/L)\delta z$. The virtual work is then, for the end mass,

$$\delta W = (f - m\ddot{z} - c\dot{z} - kz)\,\delta z \tag{1.10.1}$$

for the element of spring,

$$\delta W = [(m_s x/L^2)\,dx\,\ddot{z}][(-x/L)\,\delta z] \tag{1.10.2}$$

and for the entire spring,

$$\delta W = \left[(-m_s/L^3)\int_0^L x^2\,dx\right]\ddot{z}\,\delta z = -(m_s/3)\ddot{z}\,\delta z \tag{1.10.3}$$

Thus, for the entire system,

$$\delta W = [f - m\ddot{z} - c\dot{z} - kz - (m_s/3)\ddot{z}]\,\delta z = 0 \tag{1.10.4}$$

or

$$(m + m_s/3)\ddot{z} + c\dot{z} + kz = f \tag{1.10.5}$$

Equation (1.10.5) is of the same form as Eq. (1.6.2), except that the mass coefficient includes the end mass plus one-third of the mass of the spring. The assumed state of uniform tension in the spring is nearly correct if the mass of the spring is small compared with the mass at the end, but it errs significantly if the opposite is true.

For a second example, consider a beam shown in Fig. 1.15. The mass per unit length is m_b, the stiffness is EI, and the force per unit length is f. Both m_b and EI may vary along the length of the beam and the force f may vary along the length of the beam and also vary with time. We neglect deformation of the beam due to shear strain and we also neglect rotatory inertia, that is, the inertia of the beam elements associated with their rotation in the x–y plane. We postulate that the beam always retains some fixed shape ψ, and that only the amplitude varies with time. That is, we postulate that

$$v(x, t) = \psi(x)z(t) \tag{1.10.6}$$

The shape function $\psi(x)$ in Eq. (1.10.6) should satisfy the boundary conditions, in this case zero shear and bending moment at the free end and

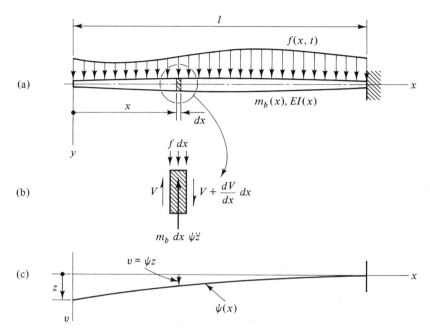

Figure 1.15 Flexible beam and assumed displacement shape. (a) Beam. (b) Forces on element. (c) Assumed displacement shape.

zero slope and deflection at the fixed end. Usually it would be chosen to have unit value at some position along the system, and the displacement variable $z(t)$ would then be the displacement at that point. We are employing the classical mathematical technique of separating variables, except that in this case, the shape function $\psi(x)$ is declared a priori instead of being derived mathematically.

We get the equation of motion by considering the dynamic translational forces on an element of the beam and then invoking the principle of virtual displacements. Figure 1.15 shows an element of the beam of length dx in its deformed position. The forces are a downward driving force, an upward inertia force, and a downward differential shear force. The displacement of the element is ψz. When we impose a virtual displacement δz, the virtual work done on the element is

$$\delta W = [f\,dx - (m_b\psi\,dx)\ddot{z} + (dV/dx)\,dx]\psi\,\delta z \qquad (1.10.7)$$

The stress–strain relations for the beam give us

$$V = -[d(EI\psi'')/dx]z \qquad (1.10.8)$$

which we differentiate to get

$$dV/dx = -[d^2(EI\psi'')/dx^2]z \qquad (1.10.9)$$

We substitute Eq. (1.10.9) into Eq. (1.10.7) and integrate over the length of the beam to get

$$\delta W = \int_0^l f\psi\,dx - \left(\int_0^l m_b\psi^2\,dx\right)\ddot{z} - \left(\int_0^l \frac{d^2(EI\psi'')}{dx^2}\psi\,dx\right)z = 0 \qquad (1.10.10)$$

The stiffness term is awkward as shown, but it can be integrated by parts using the relation

$$\int u\,dv = uv - \int v\,du$$

Let

$$u = \psi \qquad\qquad dv = [d^2(EI\psi'')/dx^2]\,dx$$
$$du = \psi'\,dx \qquad\qquad v = d(EI\psi'')/dx$$

Then

$$\int_0^l \frac{d^2(EI\psi'')}{dx^2}\psi\,dx = \left[\frac{d(EI\psi'')}{dx}\psi\right]_{x=0}^{x=l} - \int_0^l \frac{d(EI\psi'')}{dx}\psi'\,dx \qquad (1.10.11)$$

The expression in square brackets in Eq. (1.10.11) vanishes because the shear force $d(EI\psi'')/dx$ is zero at the free end and the displacement ψ is zero at the fixed end. Evaluate the remaining integral by parts again to get

$$-\int_0^l \frac{d(EI\psi'')}{dx}\psi'\,dx = -\left[EI\psi''\psi'\right]_{x=0}^{x=l} + \int_0^l EI\psi''^2\,dx \qquad (1.10.12)$$

The expression in square brackets in Eq. (1.10.12) also vanishes because the bending moment $(-EI\psi'')$ is zero at the free end and the slope (ψ') is zero at the fixed end. Thus, we finally get the equation of motion:

$$\ddot{z}\int_0^l m_b\psi^2\,dx + z\int_0^l EI\psi''^2\,dx = \int_0^l f\psi\,dx \qquad (1.10.13)$$

The system has been reduced to a linear SDF system with

$$\text{effective mass} \qquad m^E = \int_0^l m_b\psi^2\,dx \qquad (1.10.14)$$

$$\text{effective stiffness} \qquad k^E = \int_0^l EI\psi''^2\,dx \qquad (1.10.15)$$

$$\text{effective force} \qquad f^E = \int_0^l f\psi\,dx \qquad (1.10.16)$$

and the standard differential equation of motion:

$$m^E\ddot{z} + k^E z = f^E \qquad (1.10.17)$$

For a specific case, take a uniform cantilever beam of mass per unit length m_b, length l, and stiffness EI, and let the driving force per unit length $f(t)$ be uniform as well. Let us assume that the dynamic deflected shape is the same as the static deflection for a uniformly distributed load. Note that this is not exact, despite the uniformity of m, EI, and f, because the inertia force is not uniformly distributed. It is, however, a reasonable approximation and it satisfies the boundary conditions. The shape function is then

$$\psi = [(x/l)^4 - 4x/l + 3]/3 \qquad (1.10.18)$$

which has unit value at $x = 0$. The displacement coordinate z is the displacement at the zero end. Differentiating Eq. (1.10.18) twice gives

$$\psi'' = 4x^2/l^4 \qquad (1.10.19)$$

The effective mass, stiffness, and force are, respectively,

$$m^E = m_b \int_0^l \{[(x/l)^4 - 4x/l + 3]^2/9\} \, dx = \frac{104}{405} m_b l \qquad (1.10.20)$$

$$k^E = EI \int_0^l (4x^2/l^4)^2 \, dx = \frac{16EI}{5l^3} \qquad (1.10.21)$$

$$f^E = f \int_0^l \frac{[(x/l)^4 - 4x/l + 3]}{3} \, dx = \frac{2fl}{5} \qquad (1.10.22)$$

The equation of motion is, therefore,

$$\frac{104}{405} m_b l \ddot{z} + \frac{16EI}{5l^3} z = \frac{2fl}{5} \qquad (1.10.23)$$

or

$$\ddot{z} + \frac{162EI}{13 m_b l^4} z = \frac{81f}{52 m_b} \qquad (1.10.24)$$

1.11 BASE TRANSLATION

One of the dynamic problems of concern to structural engineers is the behavior of structures subjected to earthquakes. The driving force in this case is not an explicit force applied to the mass, but an implicit inertia force. Consider the simple structure of Fig. 1.16(a), which is a platform on four columns as in Fig. 1.10, but with a damper added. Let the system be driven by a dynamic ground displacement $u_g(t)$, as shown.

We define the displacement coordinate u to be the displacement of the mass *relative to the base*. Thus, the restoring force is a function of u, the

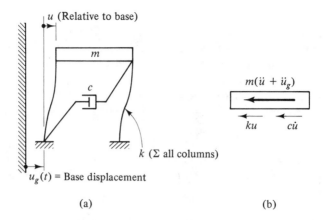

Figure 1.16 System driven by base translation. (a) System. (b) Dynamic forces.

relative displacement; the damping force is a function of $\dot{u}$, the *relative* velocity; but the inertia force is a function of $\ddot{u} + \ddot{u}_g$, the *total* acceleration. Figure 1.16(b) shows the translational forces acting on the mass, including the inertia force. Dynamic equilibrium then gives the equation of motion:

$$m\ddot{u} + c\dot{u} + ku = -m\ddot{u}_g(t) \qquad (1.11.1)$$

the same as Eq. (1.6.2) except that an inertia force $-m\ddot{u}_g(t)$ takes the place of the driving force $f(t)$.

1.12 THE VIBRATION GENERATOR

Unbalanced rotating machinery may give rise to vibrations, intentional or not. Eccentric-mass vibration generators are useful for determining experimentally the vibrational characteristics of dynamic systems. Also, most of us have experienced undesired vibrations due to such eccentric rotating masses as unbalanced tires on automobiles, broken blades on fans, and the like. The behavior is basically the same whether the vibration is intentional or not.

Figure 1.17 shows the essential components of a vibration generator. It consists of a stationary part having a mass m_S and a rotating mass m_E that rotates about a horizontal shaft with an eccentricity r and an angular frequency ω radians/sec. We show a single eccentric mass, although most real vibration generators have two or more counterrotating eccentric masses, phased so that the inertia forces augment each other in one direction and counterbalance each other in the orthogonal direction. A single eccentric

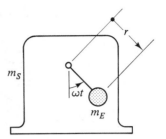

Figure 1.17   Vibration generator.

mass will suffice for the present. The horizontal displacement of the eccentric mass relative to the shaft is then $r \sin \omega t$.

The vibration generator is mounted on a structure, as shown in Fig. 1.18(a). The displacement of both the structure and the stationary part of the vibration generator is u, and the displacement of the eccentric mass is $u + r \sin \omega t$. Their accelerations are thus $\ddot{u}$ and $\ddot{u} - r\omega^2 \sin \omega t$. Figure 1.18(b) shows the forces acting on the displaced system, including the inertia forces. Dynamic equilibrium then gives the equation of motion

$$(m + m_S + m_E)\ddot{u} + c\dot{u} + ku = m_E r\omega^2 \sin \omega t \qquad (1.12.1)$$

Equation (1.12.1) is of the same form as Eq. (1.6.2), except that now the mass of both the stationary and rotating parts of the vibration generator are added to the mass of the structure in the inertia coefficient and the sinusoidal driving force depends on the rotating mass, its eccentricity, and the driving frequency. We shall return to this equation of motion in Chapter 2.

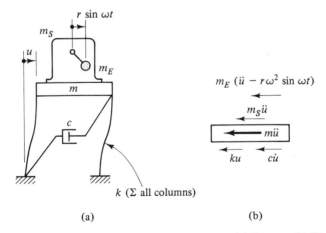

(a)   (b)

Figure 1.18   System driven by vibration generator. (a) System. (b) Dynamic equilibrium forces.

PROBLEMS

1.1 to 1.5 Write the equation of motion for the single-degree-of-freedom system shown, taking the displacement coordinate to be as indicated. All springs and dampers are weightless and all displacements are small.

1.1

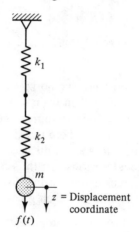

z = Displacement coordinate

1.2

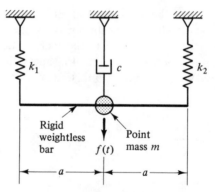

Displacement coordinate
z = Vertical displacement of mass

1.3

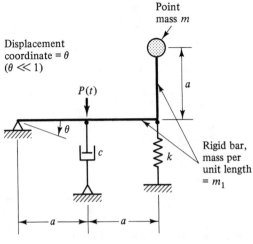

Point
mass m

Displacement
coordinate = θ
($\theta \ll 1$)

$P(t)$

θ

a

c

k

Rigid bar,
mass per
unit length
= m_1

a

a

1.4

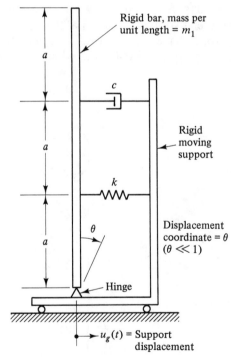

Rigid bar, mass per
unit length = m_1

a

c

Rigid
moving
support

a

k

Displacement
coordinate = θ
($\theta \ll 1$)

a

θ

Hinge

$u_g(t)$ = Support
displacement

1.5

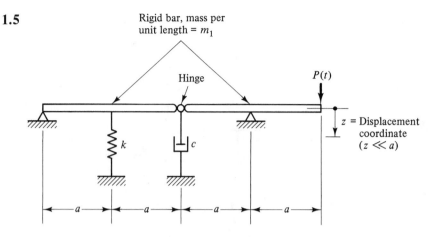

Rigid bar, mass per
unit length = m_1

Hinge

$P(t)$

k

c

z = Displacement
coordinate
$(z \ll a)$

a a a a

1.6 A rigid vehicle weighing 2000 lb, moving on a horizontal surface at a
velocity of 12 ft/sec, is stopped by a barrier consisting of wire ropes
stretched between two rigid anchors 100 ft apart, as shown in the figure.
The barrier ropes have a total cross-sectional area of 1.25 in^2 and a
modulus of elasticity of 26,000 ksi, and are stretched to an initial ten-

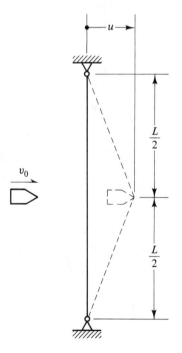

u

$\dfrac{L}{2}$

$\dfrac{L}{2}$

v_0

Plan view

sion of 1000 lb. The vehicle moves normal to the barrier and strikes it at midpoint. Find the maximum deflection of the barrier.

Assume ideal conditions, that is, a rigid vehicle, weightless barrier, no friction, no damping, and perfectly elastic wire ropes.

1.7 Write an equation of motion and supporting equations that would describe the motion of the vehicle in Problem 1.6 from the instant the vehicle strikes the barrier until maximum displacement is reached.

1.8 A uniform propped cantilever beam is subjected to a dynamic force per unit length $p(t)$, uniformly distributed along the length but varying with time. Assume that the dynamic deflected shape is the same as the static deflected shape and take the displacement coordinate to be the midspan displacement. This reduces the continuous system to a single-degree-of-freedom system. Write the differential equation of motion.

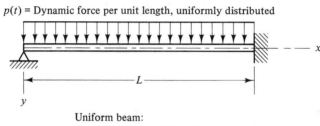

$p(t)$ = Dynamic force per unit length, uniformly distributed

Uniform beam:
m = mass per unit length
E = modulus of elasticity
I = moment of inertia

1.9 A tapered round solid cantilever beam having mass per unit volume γ and modulus of elasticity E is subjected to a dynamic force per unit length $p(t)$, uniformly distributed along its length but varying with time. Take the displacement coordinate z to be the end displacement. Assume that the deflected shape is

$$v(x, t) = \tfrac{1}{2}[3(x/L)^2 - (x/L)^3]z(t)$$

With this approximation, the continuous system is reduced to a single-degree-of-freedom system. Write the equation of motion. (Simpson's rule, with four segments, is a sufficiently accurate numerical method for evaluating the integrals involved.)

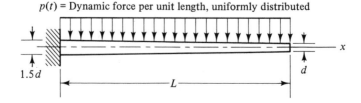

$p(t)$ = Dynamic force per unit length, uniformly distributed

CHAPTER TWO

The Linear Single-Degree-of-Freedom System

2.1 UNDAMPED FREE VIBRATION

Consider the free vibration of the undamped linear oscillator shown in Fig. 2.1, given an initial displacement u_0 and an initial velocity $\dot{u}_0$. This is an ideal system, with a rigid mass and weightless spring, and all motion is in the direction of the spring. There is no driving force.

Setting the sum of the horizontal forces, including the inertia force, equal to zero gives us the equation of motion:

$$m\ddot{u} + ku = 0 \tag{2.1.1}$$

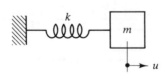

Figure 2.1 Undamped linear oscillator.

Equation (2.1.1), which is a linear homogeneous second-order differential equation with constant coefficients, has a solution of the form

$$u = e^{\lambda t} \qquad (2.1.2)$$

where there are two values of λ that will satisfy the second-order differential equation. We differentiate Eq. (2.1.2) twice to get

$$\ddot{u} = \lambda^2 e^{\lambda t} \qquad (2.1.3)$$

Now we substitute u and $\ddot{u}$ from Eqs. (2.1.2) and (2.1.3) into Eq. (2.1.1) to get

$$(m\lambda^2 + k)e^{\lambda t} = 0 \qquad (2.1.4)$$

The exponential term in Eq. (2.1.4) is never zero, so the expression in parentheses must be zero. Thus,

$$m\lambda^2 + k = 0 \qquad (2.1.5)$$

or

$$\lambda = \pm i(k/m)^{1/2} \qquad (2.1.6)$$

where $i = (-1)^{1/2}$. Thus, we obtain the two values of λ for which u from Eq. (2.1.2) will satisfy Eq. (2.1.1), and the general solution of Eq. (2.1.1) is the linear combination

$$u = C_1 exp[i(k/m)^{1/2}t] + C_2 exp[-i(k/m)^{1/2}t] \qquad (2.1.7)$$

where C_1 and C_2 are constants yet undetermined. Recognizing that the exponential function and the circular trigonometric functions are related thus:

$$\cos x = \frac{e^{ix} + e^{-ix}}{2}$$

and

$$\sin x = \frac{e^{ix} - e^{-ix}}{2i}$$

we can write Eq. (2.1.7) in a more convenient form:

$$u = A \cos (k/m)^{1/2}t + B \sin (k/m)^{1/2}t \qquad (2.1.8)$$

Equation (2.1.8) describes a sinusoidal oscillation having a frequency of $(k/m)^{1/2}$ radians per second or $(k/m)^{1/2}/2\pi$ cycles per second. We define

$$\omega_n = (k/m)^{1/2} = \text{natural frequency} \qquad (2.1.9)$$

The equation of motion, Eq. (2.1.1), then can be written

$$\ddot{u} + \omega_n^2 u = 0 \qquad (2.1.10)$$

and its solution, Eq. (2.1.8), becomes

$$u = A \cos \omega_n t + B \sin \omega_n t \qquad (2.1.11)$$

Differentiating Eq. (2.1.11), we get

$$\dot{u} = -\omega_n A \sin \omega_n t + \omega_n B \cos \omega_n t \qquad (2.1.12)$$

We evaluate Eqs. (2.1.11) and (2.1.12) at time zero to get the initial conditions:

$$u_0 = A$$
$$\dot{u}_0 = \omega_n B$$

Thus, the solution in terms of the initial conditions is

$$u = u_0 \cos \omega_n t + (\dot{u}_0/\omega_n) \sin \omega_n t \qquad (2.1.13)$$

Figure 2.2 shows the motion, plotted as displacement vs. time.

The *natural frequency* of the undamped system is ω_n, in units of radians per second. The word *frequency* usually has this meaning in technical writing, but not always. Frequency is sometimes expressed in cycles per second, or hertz, instead of radians per second, and, in that case, it is usually denoted as f instead of ω_n. The motion is periodic, repeating every $2\pi/\omega_n$ seconds, which we denote as T_n, the *natural period* of the system:

$$T_n = 2\pi/\omega_n \qquad (2.1.14)$$

The maximum displacement of the system, which we will call the *amplitude*, is $[u_0^2 + (\dot{u}_0/\omega_n)^2]^{1/2}$. While the word *amplitude* usually has this meaning in technical writing, this is not always true. Some authors call amplitude the peak-to-trough displacement. The terms *half amplitude* (meaning zero-to-peak), *amplitude* (usually meaning zero-to-peak, but occasionally meaning peak-to-trough), and *double amplitude* (meaning peak-to-trough) all appear in technical writing. In this text, we shall use the word *amplitude* consistently to mean the zero-to-peak displacement.

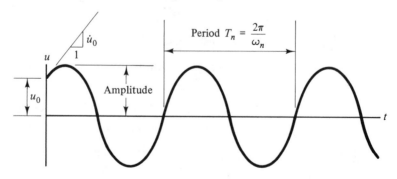

Figure 2.2   Undamped free vibration.

2.2 DAMPED FREE VIBRATION

With damping added, as in Fig. 2.3, the system's free vibration is described by the differential equation of motion:

$$m\ddot{u} + c\dot{u} + ku = 0 \qquad (2.2.1)$$

Again the solution takes the form:

$$u = e^{\lambda t} \qquad (2.1.2)$$

and again we can expect two valid values of λ. We differentiate Eq. (2.1.2) and substitute the resulting values of u, $\dot{u}$, and $\ddot{u}$ into Eq. (2.2.1) to obtain

$$(m\lambda^2 + c\lambda + k)e^{\lambda t} = 0 \qquad (2.2.2)$$

The exponential term in Eq. (2.2.2) is never zero, so the expression in parentheses must be. Hence,

$$m\lambda^2 + c\lambda + k = 0 \qquad (2.2.3)$$

or

$$\lambda = \frac{-c \pm (c^2 - 4mk)^{1/2}}{2m} \qquad (2.2.4)$$

This can also be written in the form

$$\lambda = -\frac{c}{2(km)^{1/2}}\left(\frac{k}{m}\right)^{1/2} \pm \left(\frac{k}{m}\right)^{1/2}\left(\frac{c^2}{4km} - 1\right)^{1/2} \qquad (2.2.5)$$

Let

$$\omega_n = (k/m)^{1/2} = \text{undamped natural frequency} \qquad (2.1.9)$$

and

$$\zeta = c/[2(km)^{1/2}] = c/2m\omega_n \qquad (2.2.6)$$

We call ζ the fraction of critical damping, for reasons that will appear shortly. The equation of motion, Eq. (2.2.1), can then be recast as

$$\ddot{u} + 2\zeta\omega_n\dot{u} + \omega_n^2 u = 0 \qquad (2.2.7)$$

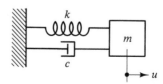

Figure 2.3 Damped linear oscillator.

and Eq. (2.2.5) becomes

$$\lambda = -\zeta\omega_n \pm \omega_n(\zeta^2 - 1)^{1/2} \qquad (2.2.8)$$

The solution of Eq. (2.2.7) is

$$u = C_1 e^{\lambda_1 t} + C_2 e^{\lambda_2 t} \qquad (2.2.9)$$

where λ_1 and λ_2 are the two values of λ from Eq. (2.2.8). The form of the solution depends upon the value of ζ.

If $\zeta > 1$, then $(\zeta^2 - 1)^{1/2}$ is real and we get

$$u = e^{-\zeta\omega_n t}\{A \exp[\omega_n(\zeta^2 - 1)^{1/2}t] + B \exp[-\omega_n(\zeta^2 - 1)^{1/2}t]\} \qquad (2.2.10)$$

If $\zeta = 1$, then $(\zeta^2 - 1)^{1/2}$ is zero (i.e., the two values of λ are the same) and Eq. (2.2.9) reduces to

$$u = e^{-\omega_n t}(A + Bt) \qquad (2.2.11)$$

If $\zeta < 1$, then $(\zeta^2 - 1)^{1/2}$ is imaginary and it is convenient to convert the exponentials into the equivalent trigonometric functions. Equation (2.2.9) then becomes

$$u = e^{-\zeta\omega_n t}\{A \cos[\omega_n(1 - \zeta^2)^{1/2}t] + B \sin[\omega_n(1 - \zeta^2)^{1/2}t]\} \qquad (2.2.12)$$

In any case, the constants A and B are determined by the values of u and $\dot{u}$ at some instant of time, usually $t = 0$.

Figure 2.4 shows the different types of response, plotted as displacement vs. time.

If $\zeta = 1$, the system returns to its equilibrium position without oscillating. This is the smallest value of the parameter ζ that inhibits oscillation completely, and it is called the critically damped case. The parameter ζ is, therefore, called the *fraction of critical damping*.

If $\zeta > 1$, the overdamped case, the system returns to the equilibrium position without oscillating, but less rapidly than in the critically damped case.

If $0 < \zeta < 1$, the system oscillates with a decaying amplitude and a frequency $\omega_n(1 - \zeta^2)^{1/2}$, somewhat less than the frequency of the undamped oscillation. This is the underdamped case, and is the one usually encountered. For this case, we define the *damped natural frequency*, in radians per second, to be

$$\omega_d = \omega_n(1 - \zeta^2)^{1/2} \qquad (2.2.13)$$

and the *damped natural period*, in seconds, is

$$T_d = 2\pi/\omega_d \qquad (2.2.14)$$

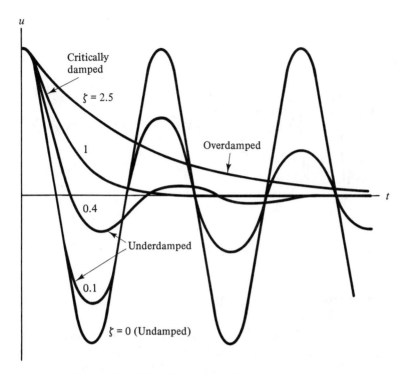

Figure 2.4   The effects of damping on free vibration.

The $\zeta = 0$ curve in Fig. 2.4 is the same undamped oscillation we had in Fig. 2.2.

The response of the underdamped case, in terms of the initial conditions, can be written as

$$u = e^{-\zeta \omega_n t}\{u_0 \cos \omega_d t + [(\dot{u}_0 + \zeta \omega_n u_0)/\omega_d] \sin \omega_d t\} \qquad (2.2.15)$$

Figure 2.5 shows several cycles of the damped free vibration of Eq. (2.2.15) for a specific case, namely, $\zeta = 0.03$.

Our initial choice of viscous damping as the energy-dissipating mechanism was mathematically convenient, for it not only provided a force that would always act to retard the motion of the system, but it also yielded equations that were easy to solve. Now we see further physical justification for the choice, because the resulting equation describes an oscillation that is constant in frequency but decays in amplitude, a sine wave inscribed in an exponential decay curve, and this is indeed quite close to the behavior we observe in most vibrating systems, at least in small-amplitude oscillations.

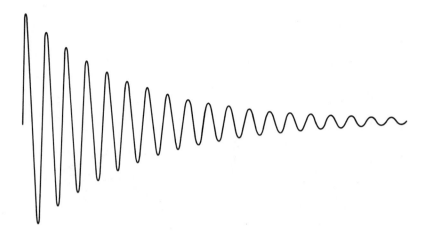

Figure 2.5   Damped free vibration.

2.3 LOGARITHMIC DECREMENT

Free vibration of a dynamic system provides one means of determining the fraction of critical damping ζ, which is one of the more elusive dynamic properties to determine. Consider the oscillation shown in Fig. 2.6.

Successive peak amplitudes are denoted as u_n, u_{n+1}, u_{n+2}, ... The peaks do not occur exactly at the point of tangency with the exponential decay curve—they precede it slightly—but the time interval between peaks

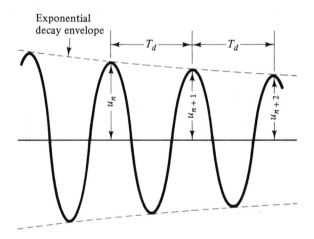

Figure 2.6   Determination of logarithmic decrement.

is a constant $T_d = 2\pi/\omega_d$. The ratio between successive peaks, u_n/u_{n+1}, is the same for all values of n. Equation (2.2.15) gives us

$$u_n/u_{n+1} = \exp[-\zeta\omega_n(t_n - t_{n+1})] = \exp[\zeta\omega_n T_d] \qquad (2.3.1)$$

The natural logarithm of this ratio is called the *logarithmic decrement,* which we designate as δ:

$$\delta = \ln(u_n/u_{n+1}) = \zeta\omega_n T_d = 2\pi\zeta/(1 - \zeta^2)^{1/2} \qquad (2.3.2)$$

from which we get

$$\zeta = \frac{\delta/2\pi}{[1 + (\delta/2\pi)^2]^{1/2}} \qquad (2.3.3)$$

or, if δ is small,

$$\zeta \approx \delta/2\pi \qquad (2.3.4)$$

Thus, the recorded free vibration of a dynamic system provides a simple basis for evaluating the fraction of critical damping.

If the decay is slow, as in Fig.2.6, it may be easier to compare amplitudes several cycles apart instead of successive amplitudes. A similar analysis gives us

$$\delta = (1/p) \ln(u_n/u_{n+p}) \qquad (2.3.5)$$

For example, suppose the oscillation in free vibration decayed from an amplitude of 0.128 in. to 0.024 in. in 18 cycles. Then, the logarithmic decrement would be

$$\delta = (1/18) \ln(0.128/0.024) = 0.0930$$

which gives

$$\zeta = 0.0148$$

Because the oscilloscope trace or other record may not have a reliable zero axis, it may be easier in practice to measure peak-to-trough double amplitudes than zero-to-peak amplitudes. It can readily be shown that the ratio of double amplitudes is the same as the ratio of zero-to-peak amplitudes, so either may be used for determining the logarithmic decrement.

2.4 UNDAMPED RESPONSE TO A BLOCK PULSE

We consider now the simplest case of forced vibration, the undamped oscillator driven by the step force shown in Fig. 2.7. The system is initially at rest, $u_0 = 0$ and $\dot{u}_0 = 0$, and the driving force jumps abruptly from zero to

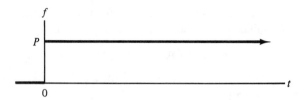

Figure 2.7 Step driving force.

a constant P at time $t = 0$. Even though the force is constant, the response is dynamic, not static.

The differential equation of motion for $t > 0$ is

$$m\ddot{u} + ku = P \qquad (2.4.1)$$

A particular solution is $u = P/k$. We add to this the general free vibration, Eq. (2.1.11), to get the general solution

$$u = P/k + A \cos \omega_n t + B \sin \omega_n t \qquad (2.4.2)$$

and its derivative

$$\dot{u} = -\omega_n A \sin \omega_n t + \omega_n B \cos \omega_n t \qquad (2.4.3)$$

We evaluate these at time zero to get

$$u(0) = P/k + A = u_0 = 0 \qquad \text{or} \qquad A = -P/k$$

and

$$\dot{u}(0) = \omega_n B = \dot{u}_0 = 0 \qquad \text{or} \qquad B = 0$$

The complete solution of Eq. (2.4.1) for $t > 0$ is thus

$$u = (P/k)(1 - \cos \omega_n t) \qquad (2.4.4)$$

After all this analysis, we arrive at what we might have deduced intuitively. The system has a new static-equilibrium position $u = P/k$, about which it oscillates as in free vibration. Figure 2.8 shows the solution.

Now modify the driving force to a block pulse, in which the force jumps from zero to P at time $t = 0$, and then drops back to zero abruptly at time $t = t_1$, as shown in Fig. 2.9. Again the system is initially at rest with zero displacement.

In the time interval $0 < t < t_1$ the motion is exactly the same as before, Eq. (2.4.4). After the pulse ends, the system is in free vibration, and we can adapt our earlier solution for undamped free vibration, Eq. (2.1.13), to the current problem just by shifting the starting time from $t = 0$ to $t = t_1$; thus,

$$u = u_1 \cos \omega_n(t - t_1) + (\dot{u}_1/\omega_n) \sin \omega_n(t - t_1) \qquad (2.4.5)$$

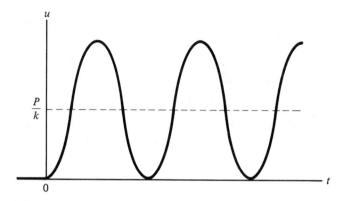

Figure 2.8 Undamped response to step driving force.

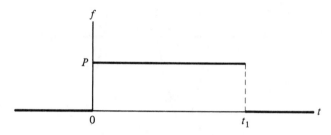

Figure 2.9 Block force pulse.

We get the values of u_1 and $\dot{u}_1$ by evaluating Eq. (2.4.4) and its derivative at time $t = t_1$:

$$u_1 = (P/k)(1 - \cos \omega_n t_1) \tag{2.4.6}$$

and

$$\dot{u}_1 = (P/k)\omega_n \sin \omega_n t_1 \tag{2.4.7}$$

We put these into Eq. (2.4.5) to get the solution:

$$\begin{aligned} u = (P/k)\{[1 - \cos \omega_n t_1] \cos \omega_n (t - t_1) \\ + \sin \omega_n t_1 \sin \omega_n (t - t_1)\} \end{aligned} \tag{2.4.8}$$

Figure 2.10 shows the solution for three different values of t_1, namely, $t_1 = T_n =$ the natural period of the oscillator, $t_1 = T_n/2$, and $t_1 = T_n/5$. The only change in the driving force in the three cases is the pulse length, yet the nature of the response varies tremendously. This is an important attribute of structural dynamics. The time characteristics of the driving force may be fully as important as its magnitude — in some cases, even more important.

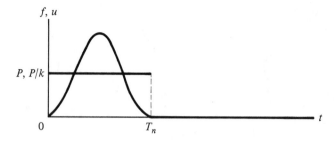

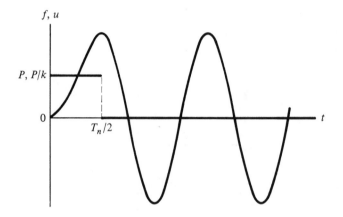

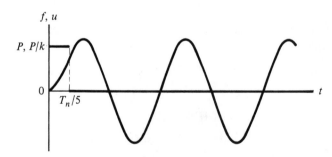

Figure 2.10 Undamped response to block pulse.

2.5 DAMPED RESPONSE TO A BLOCK PULSE

With damping in the system, the solution becomes more tedious and the expressions more cumbersome, but the process is much the same. For the step force of Fig. 2.7, the differential equation of motion is

$$m\ddot{u} + c\dot{u} + ku = P \qquad (2.5.1)$$

or

$$\ddot{u} + 2\zeta\omega_n\dot{u} + \omega_n^2 u = P/m \qquad (2.5.2)$$

Again, a particular solution is $u = P/k = P/(m\omega_n^2)$. We add to this the general solution of the free-vibration equation, which we derived in Eq. (2.2.12), to get

$$u = P/k + e^{-\zeta\omega_n t}(A \cos \omega_d t + B \sin \omega_d t) \qquad (2.5.3)$$

and its derivative

$$\begin{aligned}
\dot{u} = &-\zeta\omega_n e^{-\zeta\omega_n t}(A \cos \omega_d t + B \sin \omega_d t) \\
&+ e^{-\zeta\omega_n t}\omega_d(-A \sin \omega_d t + B \cos \omega_d t)
\end{aligned} \qquad (2.5.4)$$

Imposing the zero initial conditions, we get

$$u(0) = P/k + A = u_0 = 0 \qquad \text{or} \qquad A = -P/k$$

and

$$\dot{u}(0) = -\zeta\omega_n A + \omega_d B = \dot{u}_0 = 0 \qquad \text{or} \qquad B = -[\zeta/(1 - \zeta^2)^{1/2}]P/k$$

Thus, we get the solution for the time region $0 < t < t_1$ to be

$$u = (P/k)(1 - e^{-\zeta\omega_n t}\{\cos \omega_d t + [\zeta/(1 - \zeta^2)^{1/2}] \sin \omega_d t\}) \qquad (2.5.5)$$

Again the system has a new static equilibrium position $u = P/k$, about which it oscillates in damped free vibration. Figure 2.11 shows the solution.

After the pulse ends, the system oscillates about the zero position in damped free vibration. Again, we use the displacement and velocity at time t_1 as the initial conditions for the free-vibration phase starting at time t_1 to get, for $t > t_1$:

$$u = \exp[-\zeta\omega_n(t - t_1)]\left\{u_1 \cos \omega_d(t - t_1) + \frac{\dot{u}_1 + \zeta\omega_n u_1}{\omega_d} \sin \omega_d(t - t_1)\right\} \qquad (2.5.6)$$

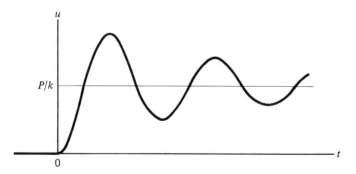

Figure 2.11 Damped response to step force.

2.6 DUHAMEL'S INTEGRAL

We can use the block-pulse solution to derive a general solution for an arbitrary driving force. First, consider the undamped oscillator driven by a short block pulse F of duration h ending at time t_1, as shown in Fig. 2.12.

We get the displacement and velocity at time t_1 from Eqs. (2.4.6) and (2.4.7):

$$u_1 = (F/k)(1 - \cos \omega_n h) \tag{2.6.1}$$

$$\dot{u}_1 = (F/k)\omega_n \sin \omega_n h \tag{2.6.2}$$

We put Eqs. (2.6.1) and (2.6.2) into Eq. (2.4.5) to get the solution:

$$\begin{aligned}u = (F/k)[&(1 - \cos \omega_n h) \cos \omega_n(t - t_1) \\ &+ \sin \omega_n h \sin \omega_n(t - t_1)]\end{aligned} \tag{2.6.3}$$

If the pulse length h is small so that $\omega_n h \ll 1$, then $\sin \omega_n h \simeq \omega_n h$ and $(1 - \cos \omega_n h) \simeq 0$, which reduces Eq. (2.6.3) to

$$u = (F/k)\omega_n h \sin \omega_n(t - t_1)$$

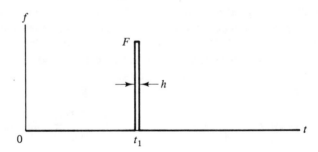

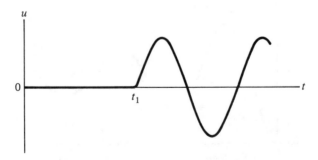

Figure 2.12 Short block pulse and undamped response.

or

$$u = \frac{Fh}{m\omega_n} \sin \omega_n(t - t_1) \qquad (2.6.4)$$

which approaches exactness as h approaches zero. Again our analysis yields a result that might be intuitively obvious. At the end of a short pulse, the velocity is the impulse divided by the mass:

$$\dot{u}_1 = Fh/m \qquad (2.6.5)$$

and the displacement is zero. Putting these conditions at time t_1 into Eq. (2.4.5), which is the equation for free vibration starting at time t_1, leads to the solution we just derived, Eq. (2.6.4).

Next consider a series of short block pulses $f_1, f_2, \ldots$ of lengths $\delta\tau_1, \delta\tau_2, \ldots$, ending at times $\tau_1, \tau_2, \ldots$, as shown in Fig. 2.13.

The effect of the ith pulse on the displacement at any time beyond the end of that pulse would be, from Eq. (2.6.4),

$$u = \frac{f_i}{m\omega_n} \delta\tau_i \sin \omega_n(t - \tau_i) \qquad (2.6.6)$$

The differential equation of motion is linear; hence, the principle of superposition is valid and we can obtain the effect of the entire train of pulses by summing the effects of the individual pulses. Thus, we obtain

$$u = \frac{1}{m\omega_n} \sum_i f_i \sin \omega_n(t - \tau_i)\,\delta\tau_i \qquad (2.6.7)$$

The summation includes all pulses up to the time t for which the displacement is sought.

The move from this to the arbitrary driving force is simple. Any continuous or piecewise continuous driving force may be approximated by a train of block pulses, as shown in Fig. 2.14. Equation (2.6.7) is valid for

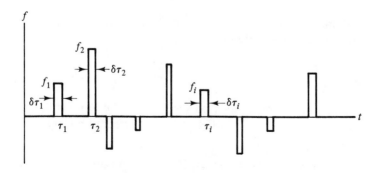

Figure 2.13 Train of short block pulses.

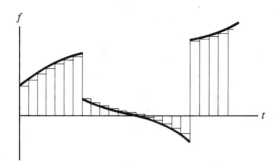

Figure 2.14 Piecewise continuous force and pulse-train approximation.

the stepwise approximation of Fig. 2.14. If we reduce the mesh size in the stepwise approximation so that the pulse lengths get shorter, in the limit as $\delta\tau_i$ tends to zero, the summation becomes an integral and the solution becomes exact:

$$u = \frac{1}{m\omega_n} \int_0^t f(\tau) \sin \omega_n(t - \tau)\, d\tau \qquad (2.6.8)$$

This is Duhamel's integral, which gives the solution for forced vibration of a single-degree-of-freedom undamped linear system for an arbitrary driving force $f(t)$ and zero initial displacement and velocity. If the initial conditions are not zero, we add to Duhamel's integral the free vibration due to the nonzero initial conditions, from Eq. (2.1.13), in order to obtain the complete solution.

Duhamel's integral for the damped system may be derived in the same manner. First, consider a short block pulse of duration h and magnitude F, ending at time t_1, as shown in Fig. 2.15. The velocity at the end of the pulse is the impulse divided by the mass:

$$\dot{u}_1 = Fh/m \qquad (2.6.5)$$

and the displacement is zero. We put these conditions at time t_1 into Eq. (2.5.6) to get the solution for $t > t_1$:

$$u = \exp[-\zeta\omega_n(t - t_1)]\frac{Fh}{m\omega_d} \sin \omega_d(t - t_1) \qquad (2.6.9)$$

For a series of short block pulses $f_1, f_2, \ldots$ of lengths $\delta\tau_1, \delta\tau_2, \ldots$, ending at times $\tau_1, \tau_2, \ldots$, as in Fig. 2.13, the effect of a single pulse is

$$u = \frac{f_i}{m\omega_d} \delta\tau_i \exp[-\zeta\omega_n(t - \tau_i)] \sin \omega_d(t - \tau_i) \qquad (2.6.10)$$

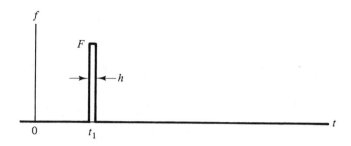

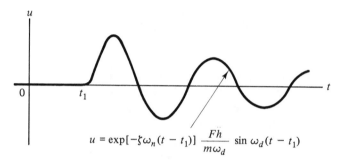

$$u = \exp[-\zeta\omega_n(t - t_1)] \frac{Fh}{m\omega_d} \sin \omega_d(t - t_1)$$

Figure 2.15   Short block pulse and damped response.

For the train of pulses, we sum the effects of the individual pulses:

$$u = \frac{1}{m\omega_d} \sum_i f_i \exp[-\zeta\omega_n(t - \tau_i)] \sin \omega_d(t - \tau_i) \delta\tau_i \qquad (2.6.11)$$

in which the summation includes all pulses prior to time t. Now if we represent an arbitrary driving force $f(t)$ as the stepwise approximation of Fig. 2.14 and reduce the mesh width, in the limit, as $\delta\tau_i$ goes to zero, the summation in Eq. (2.6.11) becomes the integral

$$u = \frac{1}{m\omega_d} \int_0^t f(\tau) \exp[-\zeta\omega_n(t - \tau)] \sin \omega_d(t - \tau) \, d\tau \qquad (2.6.12)$$

Equation (2.6.12) is Duhamel's integral for a damped system subjected to an arbitrary driving force $f(t)$ with zero initial conditions. If the initial displacement and velocity are other than zero, the free vibration that they cause, from Eq. (2.2.15), is added to Eq. (2.6.12) to obtain the complete solution.

2.7 UNDAMPED RESPONSE TO A SINUSOIDAL DRIVING FORCE

To illustrate an application of Duhamel's integral, consider first the response of an undamped linear system subjected to a sinusoidal driving force $f = P \sin \omega t$. This ω is the *driving* frequency, and is not the same as ω_n, the *natural* frequency. With this driving force, Eq. (2.6.8) becomes

$$u = \frac{1}{m\omega_n} \int_0^t [P \sin \omega\tau] \sin \omega_n(t - \tau) \, d\tau \tag{2.7.1}$$

which can also be written

$$u = \frac{P}{2m\omega_n} \int_0^t \{\cos [(\omega + \omega_n)\tau - \omega_n t] - \cos [(\omega - \omega_n)\tau + \omega_n t]\} \, d\tau \tag{2.7.2}$$

The parameter of integration is τ; t is treated as a constant during the integration. Upon integrating Eq. (2.7.2) and evaluating the result between the limits $\tau = 0$ and $\tau = t$, we obtain

$$u = \frac{P}{k} \frac{1}{1 - (\omega/\omega_n)^2} [\sin \omega t - (\omega/\omega_n) \sin \omega_n t] \tag{2.7.3}$$

Equation (2.7.3) contains two distinct vibration components, the $\sin \omega t$ term, giving an oscillation at the frequency of the driving force, and the $\sin \omega_n t$ term, giving an oscillation at the natural frequency of the system. The first of these is the *steady-state* vibration, for it is present no matter what the initial conditions are. The latter is the *transient* vibration, for it depends upon the initial conditions. The word *transient* is more accurate in the damped case, for which this component of vibration decays exponentially. If the initial displacement and velocity were

$$u_0 = 0$$

and

$$\dot{u}_0 = \frac{P}{k} \frac{\omega_n}{1 - (\omega/\omega_n)^2}$$

then the transient component would vanish, leaving only the steady-state solution:

$$u = \frac{P}{k} \frac{1}{1 - (\omega/\omega_n)^2} \sin \omega t \tag{2.7.4}$$

The factor $\{1/[1 - (\omega/\omega_n)^2]\}$ in Eq. (2.7.4) is called the *dynamic magnification factor*.

The steady-state response is a sinusoidal oscillation at the frequency of the driving force, with amplitude equal to the static displacement P/k multiplied by a dynamic magnification factor. If the frequency ratio ω/ω_n were small — that is, if the driving force oscillated slowly compared with the natural frequency of the system — then the dynamic magnification factor would be just slightly greater than 1 and the maximum displacement of the system would be essentially the same as the static displacement.

If the force oscillated just below the natural frequency of the system, the frequency ratio would be slightly less than 1, and the dynamic magnification factor would be large and positive. In that case, the vibration would be in phase with the driving force and the peak displacement would be much greater than the static displacement.

If the frequency ratio were slightly greater than 1, then the dynamic magnification factor would be large and negative. In that case, the peak displacement would be large compared with the static displacement, but the oscillation would be out of phase with the driving force — when the force acted to the right, the system would be displaced to the left.

Finally, if the driving force oscillated at a frequency far greater than the natural frequency of the system, then the dynamic magnification factor would be very small and negative, so that the system would oscillate at an amplitude much smaller than the static displacement and out of phase with the driving force.

Duhamel's integral gives the solution, but often, as in this case, it is fully as difficult to evaluate the integral as to solve the differential equation of motion some other way. The differential equation of motion for an undamped oscillator driven by a sinusoidal force is

$$m\ddot{u} + ku = P \sin \omega t \tag{2.7.5}$$

which has a particular solution of the form

$$u = A \sin \omega t \tag{2.7.6}$$

Differentiating this twice, we get

$$\ddot{u} = -\omega^2 A \sin \omega t \tag{2.7.7}$$

We put Eqs. (2.7.6) and (2.7.7) into the differential equation of motion, Eq. (2.7.5), to get

$$A = \frac{P}{k} \frac{1}{1 - (\omega/\omega_n)^2} \tag{2.7.8}$$

which we combine with Eq. (2.7.6) to get the particular solution:

$$u = \frac{P}{k} \frac{1}{1 - (\omega/\omega_n)^2} \sin \omega t \tag{2.7.4}$$

The complete solution of Eq. (2.7.5) is the particular solution, Eq. (2.7.4), and the general solution of the homogeneous (free-vibration) equation. Zero initial conditions would bring in a transient term, which would lead to Eq. (2.7.3), the same result we got from Duhamel's integral. Equation (2.7.4) is the steady-state solution, and is of major interest. We will return to it later.

2.8 DAMPED RESPONSE TO A SINUSOIDAL DRIVING FORCE

The motion of a damped linear oscillator responding to a sinusoidal driving force $P \sin \omega t$ follows the differential equation

$$m\ddot{u} + c\dot{u} + ku = P \sin \omega t \qquad (2.8.1)$$

or

$$\ddot{u} + 2\zeta\omega_n\dot{u} + \omega_n^2 u = (P/m) \sin \omega t \qquad (2.8.2)$$

A particular solution of Eq. (2.8.2) can be written in the form

$$u = C_1 \sin \omega t - C_2 \cos \omega t \qquad (2.8.3)$$

We put Eq. (2.8.3) and its first and second derivatives into Eq. (2.8.2) to get

$$[(\omega_n^2 - \omega^2)C_1 + 2\zeta\omega_n\omega C_2] \sin \omega t - [(\omega_n^2 - \omega^2)C_2 - 2\zeta\omega_n\omega C_1] \cos \omega t$$
$$= (P/m) \sin \omega t \qquad (2.8.4)$$

For Eq. (2.8.4) to be valid for all t, the coefficients of the sine terms on the two sides of the equation must be equal, and the cosine coefficients must be equal. This requirement, after we divide Eq. (2.8.4) through by ω_n^2 and use the relation $k = m\omega_n^2$, leads to the following equations for C_1 and C_2:

$$[1 - (\omega/\omega_n)^2]C_1 + 2\zeta(\omega/\omega_n)C_2 = P/k \qquad (2.8.5)$$

$$-2\zeta(\omega/\omega_n)C_1 + [1 - (\omega/\omega_n)^2]C_2 = 0 \qquad (2.8.6)$$

From Eqs. (2.8.5) and (2.8.6), we get

$$C_1 = \frac{P}{k} \frac{1 - (\omega/\omega_n)^2}{\{1 - (\omega/\omega_n)^2\}^2 + \{2\zeta\omega/\omega_n\}^2} \qquad (2.8.7)$$

$$C_2 = \frac{P}{k} \frac{2\zeta\omega/\omega_n}{\{1 - (\omega/\omega_n)^2\}^2 + \{2\zeta\omega/\omega_n\}^2} \qquad (2.8.8)$$

The oscillation of Eq. (2.8.3), with its coefficients C_1 and C_2 determined from Eqs. (2.8.7) and (2.8.8), is the steady-state response of the damped linear oscillator to a sinusoidal driving force. The motion could be displayed graphically as a plot of displacement vs. time, resulting in the familiar sine wave. Alternatively, the motion may be represented as a plot of displacement vs. velocity, the so-called *phase plane* diagram of Fig. 2.16. Displacement is plotted as the ordinate and velocity divided by ω, the frequency of the driving force, is plotted as the abscissa.

The two terms on the right side of Eq. (2.8.3) appear as rotating vectors in Fig. 2.16, the sine term as a vector of length C_1 at an angle ωt above the horizontal axis, and the cosine term as a vector of length C_2 perpendicular to the C_1 vector, that is, lagging it by a constant angle of $\pi/2$ radians. The two vectors, of unchanging length and always perpendicular, rotate counterclockwise in the phase plane at a constant angular velocity of ω radians per second.

In the geometry of Fig. 2.16, it is clear that by adding the two rotating vectors, we can represent the same motion as

$$u = (C_1^2 + C_2^2)^{1/2} \sin (\omega t - \phi) \qquad (2.8.9)$$

and

$$\dot{u}/\omega = (C_1^2 + C_2^2)^{1/2} \cos (\omega t - \phi) \qquad (2.8.10)$$

where

$$\phi = \tan^{-1}(C_2/C_1) \qquad (2.8.11)$$

If we put the values of C_1 and C_2 from Eqs. (2.8.7) and (2.8.8) into Eqs. (2.8.9) and (2.8.11) and make suitable transformations, we obtain the steady-state displacement in the form

$$u = (P/k)R_d \sin (\omega t - \phi) \qquad (2.8.12)$$

where

$$R_d = \frac{1}{\{[1 - (\omega/\omega_n)^2]^2 + [2\zeta\omega/\omega_n]^2\}^{1/2}} \qquad (2.8.13)$$

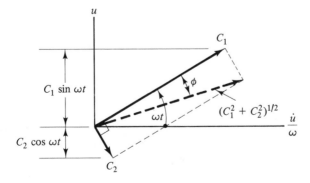

Figure 2.16   Steady-state response in phase plane.

and

$$\phi = \tan^{-1} \frac{2\zeta\omega/\omega_n}{1 - (\omega/\omega_n)^2} \qquad (2.8.14)$$

R_d is the *displacement response factor* and ϕ is the *phase lag*.

The general solution of Eq. (2.8.1) is the steady-state solution given by Eq. (2.8.12) and a transient component consisting of an exponentially decaying free vibration at the damped natural frequency of the oscillator. We now explore the terms in the steady-state solution.

2.9 DYNAMIC RESPONSE FACTORS AND PHASE LAG

The term P/k in Eq. (2.8.12) is the static displacement that would result if a constant force P were applied statically to the oscillator. The factor R_d, Eq. (2.8.13), is the *displacement response factor,* sometimes called a dynamic magnification factor, and ϕ, Eq. (2.8.14), is the *phase lag,* the amount by which the response oscillation lags behind the input force oscillation.

We differentiate the steady-state displacement to get the steady-state velocity and acceleration:

$$\dot{u} = [P/(km)^{1/2}]R_v \cos(\omega t - \phi) \qquad (2.9.1)$$

where

$$R_v = \frac{\omega/\omega_n}{\{[1 - (\omega/\omega_n)^2]^2 + [2\zeta\omega/\omega_n]^2\}^{1/2}} \qquad (2.9.2)$$

and

$$\ddot{u} = -(P/m)R_a \sin(\omega t - \phi) \qquad (2.9.3)$$

where

$$R_a = \frac{(\omega/\omega_n)^2}{\{[1 - (\omega/\omega_n)^2]^2 + [2\zeta\omega/\omega_n]^2\}^{1/2}} \qquad (2.9.4)$$

The factors R_v and R_a are the *velocity response factor* and the *acceleration response factor,* respectively. Figure 2.17 shows plots of R_d, R_v, R_a, and the phase lag ϕ as functions of the frequency ratio ω/ω_n.

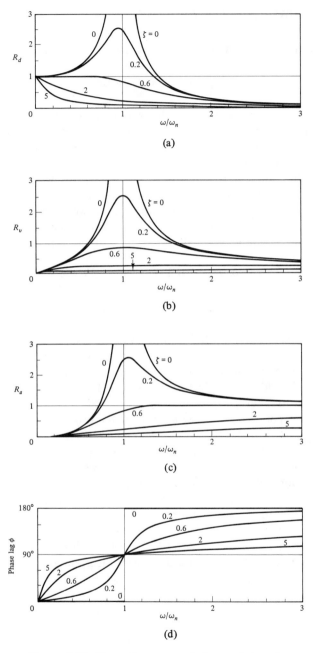

Figure 2.17 Dynamic response factors and phase lag.

2.10 THE THREE-WAY LOGARITHMIC CHART

The simple relations the dynamic response factors bear to each other, that is,

$$R_a/(\omega/\omega_n) = R_v = (\omega/\omega_n)R_d \qquad (2.10.1)$$

enable us to represent all three of them in a single three-way logarithmic chart. Because such charts will prove useful elsewhere as well, particularly in response spectrum applications to be considered in Chapter 4, we explore their construction in some detail here.

The information shown in Fig. 2.17 could equally well be shown on a log–log chart, which would show $\log R_v$ plotted as the ordinate and $\log (\omega/\omega_n)$ as the abscissa. The grid lines for R_v would be horizontal, that is, lines of constant R_v, with the R_v-axis perpendicular to them, and the ω/ω_n grid lines would be vertical grid lines, with the ω/ω_n-axis perpendicular to them.

With R_d, R_v, and ω/ω_n related as shown in Eq. (2.10.1), we can write

$$\log R_v = \log (\omega/\omega_n) + \log R_d \qquad (2.10.2)$$

If R_d were constant, this would be the equation of a straight line with a slope of $+1$. R_d grid lines would, therefore, be straight lines of slope $+1$, and the R_d-axis would be perpendicular to them. The same reasoning tells us that the R_a grid lines would be straight lines of slope -1, and the R_a-axis would be perpendicular to them.

With the point ($R_v = 1$, $\omega/\omega_n = 1$) as the origin, we draw a vertical R_v-axis and a horizontal (ω/ω_n)-axis, with equal logarithmic scales. The R_d-axis, perpendicular to the R_d grid lines, has a slope of -1, and the R_a-axis has a slope of $+1$. The logarithmic scales of the R_d and R_a axes are equal, but not the same as the R_v and ω/ω_n scales. The mark N on the R_d-axis would be located at the point ($R_v = N^{1/2}$, $\omega/\omega_n = N^{-1/2}$), and the mark N on the R_a-axis would be at the point ($R_v = N^{1/2}$, $\omega/\omega_n = N^{1/2}$, to satisfy Eq. (2.10.1). All of this leads us to the three-way logarithmic chart of Fig. 2.18, which shows R_v, R_a, and R_d for several different values of ζ, all shown as functions of ω/ω_n on a three-way logarithmic chart. The figure also shows the phase lag ϕ for the same values of ζ in a separate semi-log chart.

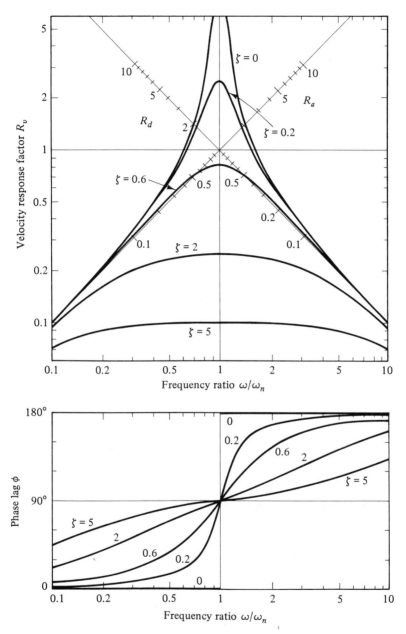

Figure 2.18   Three-way logarithmic chart of dynamic response factors and phase lag.

2.11 THE VIBRATION GENERATOR

The equation of motion for a single-degree-of-freedom damped system driven by a vibration generator is

$$m\ddot{u} + c\dot{u} + ku = m_E r\omega^2 \sin \omega t \qquad (2.11.1)$$

in which the mass term m must include the mass of the vibration generator. The steady-state solution is

$$u = (m_E/m)R_a r \sin (\omega t - \phi) \qquad (2.11.2)$$

The *displacement* is the mass ratio times the *acceleration* response factor R_a times the eccentricity of the rotating mass, with a phase lag ϕ.

2.12 BASE TRANSLATION

If a sinusoidal base translation $u_g = A \sin \omega t$ drives the system instead of a sinusoidal force $P \sin \omega t$, the equation of motion is

$$m\ddot{u} + c\dot{u} + ku = -m\ddot{u}_g = m\omega^2 A \sin \omega t \qquad (2.12.1)$$

or

$$\ddot{u} + 2\zeta\omega_n\dot{u} + \omega_n^2 u = A\omega^2 \sin \omega t \qquad (2.12.2)$$

The steady-state solution is

$$u = AR_a \sin (\omega t - \phi) \qquad (2.12.3)$$

The *displacement* is equal to the base displacement multiplied by the *acceleration* response factor R_a, with a phase lag ϕ.

2.13 DAMPING FROM RESONANCE CURVES

Resonance curves provide an alternative method of evaluating the damping coefficient ζ. Figure 2.19 represents an experimental response curve for a driving force $P \sin \omega t$. A_r is the maximum amplitude of steady-state displacement achieved as the driving frequency sweeps past resonance, and ω_1 and ω_2 are driving frequencies on either side of resonance at which the steady-state amplitude is $A = A_r/n$. At the peak, the value of R_d is $1/2\zeta(1 - \zeta^2)^{1/2}$, at a driving frequency just slightly lower than ω_n, and R_d has a value $1/n$ times its peak value at driving frequencies ω_1 and ω_2. From Eq. (2.8.13), we conclude that both of these driving frequencies satisfy the equation

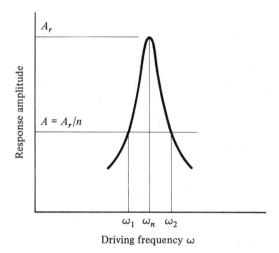

Figure 2.19 Determining damping from resonance curve.

$$\frac{1}{\{[1 - (\omega/\omega_n)^2]^2 + (2\zeta\omega/\omega_n)^2\}^{1/2}} = \frac{1}{2n\zeta(1 - \zeta^2)^{1/2}} \qquad (2.13.1)$$

We invert both sides of Eq. (2.13.1), square them, and rearrange terms to get

$$(\omega/\omega_n)^4 - 2(1 - 2\zeta^2)(\omega/\omega_n)^2 + 1 - 4n^2\zeta^2(1 - \zeta^2) = 0 \qquad (2.13.2)$$

Equation (2.13.2) is a quadratic equation in $(\omega/\omega_n)^2$, the roots of which are

$$(\omega/\omega_n)^2 = 1 - 2\zeta^2 \pm 2\zeta[(n^2 - 1)(1 - \zeta^2)]^{1/2} \qquad (2.13.3)$$

If now we neglect ζ^2 in comparison with 1 in the last term of Eq. (2.13.3) and subtract the smaller root from the larger one, we get

$$\frac{\omega_2^2 - \omega_1^2}{\omega_n^2} = 4\zeta(n^2 - 1)^{1/2} \qquad (2.13.4)$$

Finally, if we use the approximation

$$\omega_n \simeq (\omega_1 + \omega_2)/2 \qquad (2.13.5)$$

then Eq. (2.13.4) reduces to

$$\zeta = \frac{\omega_2 - \omega_1}{\omega_2 + \omega_1} \frac{1}{(n^2 - 1)^{1/2}} \qquad (2.13.6)$$

or, if the driving frequency is in hertz instead of radians/second,

$$\zeta = \frac{f_2 - f_1}{f_2 + f_1} \frac{1}{(n^2 - 1)^{1/2}} \qquad (2.13.7)$$

For example, suppose that Fig. 2.19 portrayed the experimental response of a system to a sinusoidal input force of constant amplitude and variable frequency. Further, suppose that the peak response amplitude was 0.630 in, and that on either side of resonance at frequencies of 10.9 hertz and 13.0 hertz, the response amplitude was 0.333 in. For this case, we would have

$$n = 0.630/0.333 = 1.89$$

and the damping coefficient, from Eq. (2.13.7), would be approximately

$$\zeta = \frac{13.0 - 10.9}{13.0 + 10.9} \frac{1}{[(1.89)^2 - 1]^{1/2}} = 0.055$$

If the system were driven by a vibration generator instead of a constant-amplitude sinusoidal force, Eq. (2.13.7) could still be used to calculate ζ, but the approximation would be coarser. In either case, the approximation is better if damping is small, that is, if the response curve has a sharp spike at resonance.

2.14 THE SEISMIC PICKUP

The seismic pickup is a special linear SDF system—essentially a mass–spring–damper system constrained to displace in only one direction. Actually, many seismic instruments contain three separate pickups oriented to respond in three orthogonal directions. The instrument is attached to the structure whose motion is to be recorded, and the pickup itself is then a damped linear SDF system driven by its base motion—the motion of the supporting structure.

Figure 2.20 shows the essential components schematically. The mass of the pickup, the pendulum, vibrates as the supporting structure moves. The displacement of the pendulum relative to the base of the pickup, u, is magnified by a factor L, often by optical means, and recorded, perhaps on film or on magnetic tape, or possibly even by means of a stylus trace on a smoked cylinder. Whatever the form of the record, the end result is a trace showing the displacement of the pendulum relative to the base of the instrument as a function of time, to some scale. The magnification factor may be small or large, depending on what the instrument is designed to accomplish.

Figure 2.21 shows part of the strong-motion accelerogram of the main shock of the San Fernando earthquake of February 9, 1971, recorded in the epicentral region near Pacoima Dam of the Los Angeles County Flood District. The instrument is part of the accelerograph network of the National Oceanographic and Atmospheric Administration. The record shows three

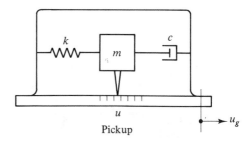

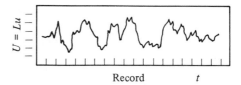

Figure 2.20 The essential components of a seismic instrument.

components of ground acceleration, along with time traces at the top and bottom of the chart.

First, we look at the relation between the seismograph record and the motion that induced it from a primitive and somewhat intuitive viewpoint. The differential equation of motion for the pickup is

$$\ddot{u} + 2\zeta\omega_n\dot{u} + \omega_n^2 u = -\ddot{u}_g(t) \qquad (2.14.1)$$

We integrate Eq. (2.14.1) twice to get

$$u + 2\zeta\omega_n \int_0^t u\, d\tau + \omega_n^2 \int_0^t \int_0^\tau u\, d\tau'\, d\tau = -u_g \qquad (2.14.2)$$

or

$$u = -u_g - 2\zeta\omega_n \int_0^t u\, d\tau - \omega_n^2 \int_0^t \int_0^\tau u\, d\tau'\, d\tau \qquad (2.14.3)$$

If ω_n is small enough to make the integral terms in Eq. (2.14.3) negligible, then the pickup displacement u is the negative of the base displacement u_g. In that case, the pickup acts as a displacement meter. This suggests that if the pickup is to record displacements, it must have a low natural frequency — a long natural period. This implies either large mass or small stiffness or both, and all would create difficulties in handling and transporting the instrument.

These results are not entirely satisfying, for we spoke of ω_n being small without defining *small*. Also, the two error terms that the "small" ω_n was to nullify in Eq. (2.14.3) are integrals, which may grow with time. If we were to put the instrument on a truck and drive to the next county, we could hardly expect it to record the displacement of the truck. But at least

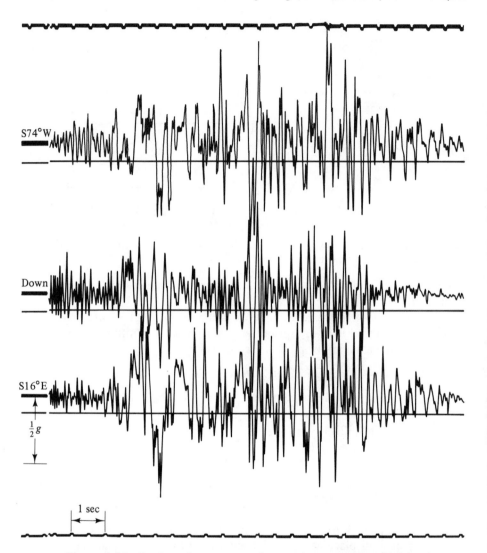

Figure 2.21 Portion of a strong-motion accelerogram. (Provided by the Earthquake Engineering Research Laboratory, California Institute of Technology.)

our look at the equations suggests that a displacement meter must be a low-frequency instrument. Intuitively, this makes sense — with extreme flexibility the mass would tend to stand still while the base moved.

We can also write the equation of motion for the pickup as

$$u = (1/\omega_n^2)(-\ddot{u}_g - \ddot{u} - 2\zeta\omega_n\dot{u}) \qquad (2.14.4)$$

The trace written on the record, U, is this displacement u magnified by a lever factor L:

$$U = Lu = -(L/\omega_n^2)\ddot{u}_g - (1/\omega_n^2)\ddot{U} - (2\zeta/\omega_n)\dot{U} \qquad (2.14.5)$$

If ω_n and the lever factor L are both large, the last two terms in Eq. (2.14.5) will be small, but the first one will not, and we will have, approximately,

$$U \simeq -(L/\omega_n^2)\ddot{u}_g \qquad (2.14.6)$$

Thus, if ω_n is large, the trace amplitude U will be nearly proportional to the base acceleration, and, therefore, the pickup will act as an accelerometer.

    This, too, is less than satisfying, for *large* has not been defined, and the terms in Eq. (2.14.5) to be nullified by the "large" ω_n are $\dot{U}$, the slope of the trace, and $\ddot{U}$, the curvature of the trace, either of which may be quite large indeed. So what we have acquired to this point is a rough indication that an accelerometer must be a high-frequency instrument and a displacement meter must be a low-frequency instrument.

    We now explore more fully the relation between the record and the motion that caused it. Because any arbitrary base motion can be resolved into sinusoidal components by means of a Fourier series or the Fourier transform, we may consider a sinusoidal base displacement and then infer from the results what the behavior would be for a more general base motion. Let the base displacement be

$$u_g = A \sin \omega t \qquad (2.14.7)$$

The equation of motion of the pickup is then

$$\ddot{u} + 2\zeta\omega_n\dot{u} + \omega_n^2 u = -\ddot{u}_g = \omega^2 A \sin \omega t \qquad (2.14.8)$$

The response of the pickup is

$$u(t) = AR_a \sin (\omega t - \phi) = R_a u_g(t - \phi/\omega) \qquad (2.14.9)$$

The relative displacement of the pendulum at time t is R_a times what the base displacement was at time $t - \phi/\omega$. The pendulum displacement is the base displacement modified by a factor R_a and a time lag of ϕ/ω. If the factor R_a were a constant, the instrument would behave as a displacement meter, for the ordinate of the trace would correspond to the base displacement, except for the time shift. Figures 2.17 and 2.18 show that R_a approaches 1 as the frequency ratio approaches infinity. Moreover, in the curve for $\zeta = 0.6$ in Fig. 2.18, we see that R_a is nearly equal to 1 for all frequency ratios greater than 1. We therefore conclude that a seismic pickup damped to about 60 percent of critical damping will record displacements for input motion of higher frequency (i.e., of shorter period) than that of the pickup. The displacement meter must be of lower frequency (longer period) than the input motion it is to record.

From Eq. (2.10.1) we have $R_a = (\omega/\omega_n)^2 R_d$, which enables us to rewrite Eq. (2.14.9) as

$$u(t) = (1/\omega_n^2)R_d A\omega^2 \sin(\omega t - \phi) \qquad (2.14.10)$$

or

$$u(t) = -(1/\omega_n^2)R_d \ddot{u}_g(t - \phi/\omega) \qquad (2.14.11)$$

The relative displacement of the pendulum at time t is $-R_d/\omega_n^2$ times what the base acceleration was at time $t - \phi/\omega$. The recorded trace is the base acceleration modified by a factor $-LR_d/\omega_n^2$ and recorded with a time lag ϕ/ω. Because L and ω_n are instrument constants independent of the base motion, if the factor R_d were a constant, the instrument would record accelerations. Figures 2.17 and 2.18 show that R_d approaches 1 as ω/ω_n approaches zero. Moreover, Fig. 2.18 shows that for $\zeta = 0.6$, R_d is close to 1 for all ω/ω_n less than 1. We conclude that a seismic pickup damped to about 60 percent of critical damping will respond as an accelerometer to base motion lower in frequency than the frequency of the instrument.

A few observations are significant. First, the high-frequency instrument that serves as an accelerometer must have a small mass or be very stiff or both. It can, therefore, be rugged and can be built to withstand rough handling without distress. Second, a high-frequency ω_n requires a large magnification factor L to get a readable record, and the noise is magnified as well as the signal.

The phase lag is

$$\phi = \tan^{-1}\frac{2\zeta\omega/\omega_n}{1 - (\omega/\omega_n)^2} \qquad (2.8.14)$$

and if ω/ω_n is small,

$$\phi \simeq 2\zeta\omega/\omega_n \qquad (2.14.12)$$

or

$$\phi/\omega \simeq 2\zeta/\omega_n = \text{constant} \qquad (2.14.13)$$

Accelerometers operate in the range of small ω/ω_n, and, therefore, the time shift is the same for all frequency components of base motion. The accelerometer, therefore, writes an undistorted record.

PROBLEMS

2.1 The ideal single-degree-of-freedom system shown below, having frequency ω_n and negligibly small damping, is subjected to a sinusoidal base displacement of constant frequency ω and constant amplitude A.

  For what range of frequency ratio ω/ω_n will the amplitude of the total steady-state displacement (i.e., $u + u_g$) be less than twice the amplitude of the base displacement?

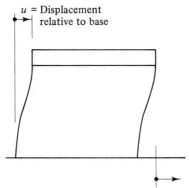

u = Displacement relative to base

Base displacement $u_g = A \sin \omega t$

2.2 A vertical cantilever, a 3 in × 3 in × ³⁄₁₆ in tube 60 in long, supports a 2000-lb weight attached at the tip, as shown in the figure. The properties of the tube are

$$A = 2.02 \text{ in}^2$$
$$S = 1.73 \text{ in}^3$$
$$I = 2.60 \text{ in}^4$$
$$E = 29,500 \text{ ksi}$$

The system is subjected to a sinusoidal force at the tip, acting horizontally in one of the planes of symmetry. The force has an amplitude of

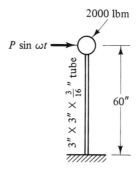

2000 lbm

$P \sin \omega t$ →

3" × 3" × $\frac{3}{16}$" tube

60"

250 lb and oscillates at 3 cycles per second. Assuming that the system is damped to 2 percent of critical damping, find the maximum steady-state bending stress in the cantilever.

Treat the attached weight as a point mass and neglect the weight of the tube. Neglect $P-\Delta$ effects, that is, neglect the bending moment due to the eccentricity of the gravity force on the tip mass with respect to the base of the cantilever.

2.3 Damping in seismic instruments is sometimes expressed as a damping ratio D, the ratio of peak-to-trough amplitudes of free vibration one-half cycle apart, that is, $D = a/b$ in the curve.

    (a) Derive the relation between the fraction of critical damping and the damping ratio D.

    (b) Damping ratios of about 9 to 11 are typical of strong-motion accelerographs. What are the corresponding fractions of critical damping?

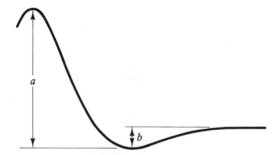

2.4 A rigid beam, 6 ft long and weighing 10 lb/ft, rests on a fulcrum 2 ft from one end and is supported by a weightless spring of stiffness 720 lb/ft at one end. At the other end is a viscous damper, $c = 0.30$ lb/(ft/sec). The beam is driven by a force that is uniformly distributed along the length of the beam and that varies sinusoidally with time, having an amplitude of 12 lb/ft and a frequency of 1.5 cycles per second.

    Find the amplitude of the steady-state displacement.

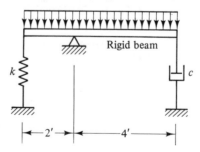

2.5 Consider two ideal single-degree-of-freedom undamped systems, (a) a spring–mass system having a mass of 100 lbm (3.11 slugs) attached to the end of a spring of stiffness 25 lb/ft, and (b) a pendulum having a mass of 100 lbm (3.11 slugs) and a length of 4 ft. In free vibration at small amplitude on earth, where the acceleration of gravity is 32.174 ft/sec², both systems have periods of 2.22 sec.

Suppose both systems were transported to the moon, where the acceleration of gravity is 5.5 ft/sec². What would be their periods of free vibration there?

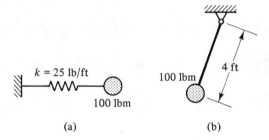

$k = 25$ lb/ft   100 lbm   4 ft

100 lbm

(a)   (b)

CHAPTER THREE

Numerical Methods for Solving the Differential Equations of Motion

3.1 THE EQUATIONS OF MOTION

In structural dynamics, we often work with linear differential equations of motion, such as the classical

$$\ddot{u} + 2\zeta\omega_n\dot{u} + \omega_n^2 u = f(t)/m$$

which yield readily to analytical methods of solution. With numerical methods, however, we can just as easily consider a more general case, in which the acceleration $\ddot{u}$ is some function of displacement, velocity, and time:

$$\ddot{u} = f(u, \dot{u}, t) \qquad (3.1.1)$$

where f denotes simply some function of the indicated variables, not a driving force. The function may be linear or nonlinear, simple or complex, and may involve the histories of the variables as well as their current values, as would be the case for hysteretic systems.

Numerical methods are also applicable to multidegree-of-freedom systems, which we will consider in Chapter 5.

3.2 TAYLOR SERIES

Numerical processes for solving differential equations are almost countless. We will consider a few that are attractive for solving the differential equations of motion. To approach any of them, it will be useful to consider the Taylor series.

A mathematical function that is continuous and has continuous derivatives of all orders is an **analytic** function. Examples are the polynomials and the circular and hyperbolic sines and cosines. Any arbitrary piecewise continuous function, that is, a function that has at most a finite number of discontinuities, can be approximated as closely as we wish by a superposition of analytic functions. For example, Fourier series can be used to represent any arbitrary piecewise continuous function as a series of sines and cosines, or the Legendre polynomials to represent it as a polynomial.

If the value of an analytic function $u(t)$ and all its derivatives were known for one time t_0, then we could compute the value of the function for any other time $t_0 + h$ from the Taylor series; thus,

$$u(t_0 + h) = u(t_0) + h\dot{u}(t_0) + (h^2/2)\ddot{u}(t_0)$$
$$+ (h^3/3!)\dddot{u}(t_0) + \ldots + (h^n/n!)u^{(n)}(t_0) + \ldots \tag{3.2.1}$$

In principle, the projection interval h could be any size, great or small, for the factorial in the denominator would eventually dominate and cause all following terms to be negligibly small. In practice, we need a small interval h to make the series converge rapidly. To abbreviate the notation, let

$$u_n = \text{displacement at time } t_n$$
$$\ddot{u}_{n+1} = \text{acceleration at time } t_{n+1} = t_n + h$$
$$\dot{u}_{n-3} = \text{velocity at time } t_{n-3} = t_n - 3h, \text{ etc.}$$

The Taylor series expansion about time t_n gives

$$u_{n+1} = u_n + h\dot{u}_n + (h^2/2)\ddot{u}_n + (h^3/6)\dddot{u}_n + (h^4/24)u_n^{iv} + \ldots \tag{3.2.2}$$

3.3 THE LINEAR-ACCELERATION METHOD

We differentiate Eq. (3.2.2) term by term to get

$$\dot{u}_{n+1} = \dot{u}_n + h\ddot{u}_n + (h^2/2)\dddot{u}_n + (h^3/6)u_n^{iv} + \ldots \tag{3.3.1}$$

and

$$\ddot{u}_{n+1} = \ddot{u}_n + h\dddot{u}_n + (h^2/2)u_n^{iv} + \ldots \tag{3.3.2}$$

We get $\ddot{u}_n$ from Eq. (3.3.2) and insert it into Eqs. (3.2.2) and (3.3.1) to get

$$u_{n+1} = u_n + h\dot{u}_n + (h^2/6)(2\ddot{u}_n + \ddot{u}_{n+1}) - (1/24)h^4 u_n^{iv} + \dots \qquad (3.3.3)$$

$$\dot{u}_{n+1} = \dot{u}_n + (h/2)(\ddot{u}_n + \ddot{u}_{n+1}) - (1/12)h^3 u_n^{iv} + \dots \qquad (3.3.4)$$

The final terms shown in Eqs. (3.3.3) and (3.3.4) are the dominant error terms. We discard them and succeeding terms in the series to get the recurrence relations for the linear-acceleration method. We use them in the following manner.

Knowing the displacement u_n and velocity $\dot{u}_n$ at some time t_n (ordinarily, the initial conditions would be the starting point), we compute $\ddot{u}_n$ from the differential equation of motion, Eq. (3.1.1). We choose a time interval h and then:

1. Estimate $\ddot{u}_{n+1}^*$, the acceleration at the end of the interval. The beginning acceleration $\ddot{u}_n$ would be the simplest first estimate of $\ddot{u}_{n+1}^*$.

2. Compute

$$u_{n+1} = u_n + h\dot{u}_n + (h^2/6)(2\ddot{u}_n + \ddot{u}_{n+1}^*) \qquad (3.3.5)$$

3. Compute

$$\dot{u}_{n+1} = \dot{u}_n + (h/2)(\ddot{u}_n + \ddot{u}_{n+1}^*) \qquad (3.3.6)$$

4. Compute

$$\ddot{u}_{n+1} = f(u_{n+1}, \dot{u}_{n+1}, t_{n+1}) \qquad (3.1.1)$$

5. (a) If $\ddot{u}_{n+1} \neq \ddot{u}_{n+1}^*$, take the computed $\ddot{u}_{n+1}$ as an improved estimate $\ddot{u}_{n+1}^*$ and return to step 2.

   (b) If $\ddot{u}_{n+1} = \ddot{u}_{n+1}^*$, the iteration has converged. Advance to the next time interval and return to step 1.

If the system is undamped, so that the velocity term is absent from Eq. (3.1.1), then the velocity $\dot{u}_{n+1}$ is not needed until the next time step and we may defer Step 3 until the iteration has converged.

To illustrate the process, we carry out several steps of the calculations for a damped linear oscillator in free vibration. The differential equation of motion is

$$\ddot{u} = -2\zeta\omega_n\dot{u} - \omega_n^2 u$$

We use an oscillator with period $T_n = 1$ sec and damping $\zeta = 0.05$, with initial displacement $u_0 = 1$ in and initial velocity $\dot{u}_0 = 0$. We use a time step $h = 0.125$ sec, one-eighth of the natural period. For the present, take this to be an arbitrary choice, for we have not yet considered a proper choice of the time step.

TABLE 3.1 Linear Acceleration Calculations

$T_n = 1.00$ sec			$u_0 = 1.00$ in		$\omega_n = 6.283$ rad/sec		
$\zeta = 0.05$			$\dot{u}_0 = 0.00$		$h = 0.125$ sec		
1	2	3	4	5	6	7	8
t (sec)	$u_n + h\dot{u}_n$ $+ h^2\ddot{u}_n/3$ (in)	$h^2\ddot{u}/6$ (in)	u (in)	$\dot{u}_n + h\ddot{u}_n/2$ (in/sec)	$h\ddot{u}/2$ (in/sec)	$\dot{u}$ (in/sec)	$\ddot{u}$ (in/sec^2)
0.000			1.0000			0.000	−39.48
0.125	0.7944	−0.1028	0.6916	−2.467	−2.467	−4.935	−24.20
		−0.0630	0.7314		−1.513	−3.980	−26.37
		−0.0687	0.7257		−1.648	−4.116	−26.06
		−0.0679	0.7265		−1.629	−4.096	−26.11
		−0.0680	0.7264		−1.632	−4.099	−26.10
0.250	0.0781	−0.0680	0.0101	−5.730	−1.631	−7.362	4.23
		0.0110	0.0891		0.264	−5.466	−0.08
		−0.0002	0.0778		−0.005	−5.736	0.53
		0.0014	0.0794		0.033	−5.697	0.44
		0.0012	0.0792		0.028	−5.703	0.46
0.375	−0.6313	0.0012	−0.6301	−5.674	0.028	−5.646	28.42
		0.0740	−0.5572		1.776	−3.898	24.45
		0.0637	−0.5676		1.528	−4.146	25.01
		0.0651	−0.5661		1.563	−4.111	24.93
		0.0649	−0.5663		1.558	−4.116	24.94
0.500	−0.9509	0.0650	−0.8859	−2.557	1.559	−0.998	35.60
		0.0927	−0.8582		2.225	−0.332	34.09
		0.0888	−0.8621		2.131	−0.426	34.30
		0.0893	−0.8616		2.144	−0.413	34.27
		0.0893	−0.8617		2.142	−0.415	34.28
0.625	−0.7350	0.0893	−0.6457	1.727	2.142	3.870	23.06
		0.0601	−0.6749		1.441	3.169	24.65
		0.0642	−0.6708		1.541	3.268	24.43
		0.0636	−0.6714		1.527	3.254	24.46
		0.0637	−0.6713		1.529	3.256	24.46

Table 3.1 shows the linear-acceleration method calculations. The first row shows the given initial displacement and velocity, and the initial acceleration determined from the differential equation. In each time step thereafter, we enter t in the first column, and in column 2 the part of Eq. (3.3.5) that involves conditions at the start of the interval. We begin the time step by assuming that the acceleration $\ddot{u}^*$ at the end of the interval will be the same as $\ddot{u}$ at the beginning. We compute $h^2\ddot{u}^*/6$, the remaining term of Eq. (3.3.5), in column 3, and then add columns 2 and 3 to get the end displacement in column 4. Similarly, we enter in column 5 the part of Eq. (3.3.6) that involves conditions at the start of the interval, compute $h\ddot{u}^*/2$, the remaining term of Eq. (3.3.6), in column 6, and then add

columns 5 and 6 to get the end velocity in column 7. Now we use the end displacement and velocity, columns 4 and 7, respectively, in the equation of motion to recompute the end acceleration, column 8. This differs from the previous $\ddot{u}^*$, so we use it as a new estimate $\ddot{u}^*$ and repeat the calculations of columns 3, 4, 6, 7, and 8. Columns 1, 2, and 5 remain unchanged until the next time step. We continue the iteration until the difference between two successive values of $\ddot{u}^*$ in column 8 is negligible. When this is achieved, we advance to the next time step and run through the whole process again.

Error is generated in the numerical process, of course. Figure 3.1 compares the computed displacements with the exact solution for the first few cycles of oscillation. Open circles show the linear-acceleration-method results. The deviation from the exact solution seems to be as much a phase shift as a deviation in amplitude, the numerical solution lagging behind the exact solution. Later in this chapter, we will take a more detailed look at error and how the choice of time step h affects error in this and other numerical processes.

The linear-acceleration method is a **single-step** process, for each time step employs only information pertaining to that time interval $t_n \leq t \leq t_{n+1}$. Earlier information or future information is not used. This is advantageous because it permits altering the time interval h whenever that is necessary or desirable during the solution.

The method is **iterative** because the acceleration at the end of the interval appears in the recurrence relations. That acceleration affects the computed displacement and velocity, which in turn determine the acceleration that was used to compute them. This requires that a sequence of computations be repeated until it converges to a solution.

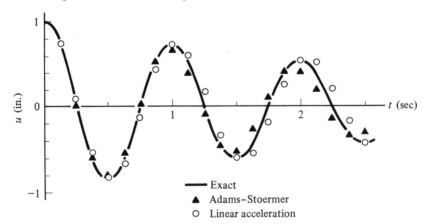

Figure 3.1   Comparison of numerical and exact solutions.

The process is nominally **third order** because all powers of h through the third are taken into account in the displacement-recurrence relation, Eq. (3.3.5). The dominant error term deleted from Eq. (3.3.3) is of order h^4. The velocity-recurrence relation, Eq. (3.3.6), is second order, but the velocity is multiplied by h when it is used for computing displacement in the next time step.

The recurrence relations account for all derivatives through the third in the Taylor series expansion; the error terms discarded from Eqs. (3.3.3) and (3.3.4) all contain the fourth and higher derivatives. If the acceleration were truly linear throughout the time interval, the third derivative would be a constant and the fourth and higher derivatives would all be zero. In those circumstances, the method would be exact—hence the name **linear-acceleration** method.

As with most iterative processes, circumstances could arise that would make the iteration diverge. For the linear oscillator, the iteration will converge if the time interval h is less than 0.39 times the natural period of the system. Convergence, however, does not imply accuracy. Reasonable accuracy requires a smaller h. A value of h as great as one-fifth of the natural period might give acceptable results for the first cycle or two of oscillation. Something of the order of $T_n/10$ is often used for hand calculations, and as small as $T_n/50$ to $T_n/100$ with computers. A smaller h leads to more rapid convergence, of course, so the larger number of time steps is at least partially offset by a smaller number of iterations in each step. In any case, h must be small enough to let the driving force be portrayed accurately.

3.4 THE ADAMS–STOERMER METHOD

The Adams–Stoermer method, surely among the simplest and fastest numerical methods, achieves the same order of accuracy as the linear-acceleration method without iteration. We can derive a version of the Adams–Stoermer method from forward and backward Taylor series expansions about t_n:

$$u_{n+1} = u_n + h\dot{u}_n + (h^2/2)\ddot{u}_n + (h^3/3!)\dddot{u}_n + (h^4/4!)u_n^{iv} + \ldots \qquad (3.4.1)$$

$$u_{n-1} = u_n - h\dot{u}_n + (h^2/2)\ddot{u}_n - (h^3/3!)\dddot{u}_n + (h^4/4!)u_n^{iv} + \ldots \qquad (3.4.2)$$

Adding these and rearranging terms, we get

$$u_{n+1} = 2u_n - u_{n-1} + h^2\ddot{u}_n + (1/12)h^4 u_n^{iv} + \ldots \qquad (3.4.3)$$

Differentiating Eq. (3.4.1) gives us

$$\dot{u}_{n+1} = \dot{u}_n + h\ddot{u}_n + (h^2/2)\dddot{u}_n + (h^3/3!)u_n^{iv} + \ldots \qquad (3.4.4)$$

and differentiating Eq. (3.4.2) twice gives us

$$\ddot{u}_{n-1} = \ddot{u}_n - h\dddot{u}_n + (h^2/2)u_n^{iv} + \ldots \tag{3.4.5}$$

We find $\dddot{u}_n$ from Eq. (3.4.5) and substitute that in Eq. (3.4.4) to get

$$\dot{u}_{n+1} = \dot{u}_n + (h/2)(3\ddot{u}_n - \ddot{u}_{n-1}) + (5/12)h^3u_n^{iv} + \ldots \tag{3.4.6}$$

We discard the error terms from Eqs. (3.4.3) and (3.4.6) to get the desired recurrence relations. One complete time step for the Adams–Stoermer process comprises:

1. Compute

$$u_{n+1} = 2u_n - u_{n-1} + h^2\ddot{u}_n \tag{3.4.7}$$

2. Compute

$$\dot{u}_{n+1} = \dot{u}_n + (h/2)(3\ddot{u}_n - \ddot{u}_{n-1}) \tag{3.4.8}$$

3. Compute

$$\ddot{u}_{n+1} = f(u_{n+1}, \dot{u}_{n+1}, t_{n+1}) \tag{3.1.1}$$

Table 3.2 shows the Adams–Stoermer calculations for the example of Table 3.1, a damped linear oscillator in free vibration. The first line of the table shows the given initial displacement and velocity and the initial acceleration computed from the differential equation. Immediately, an obstacle appears. To apply the numerical process, we need two past values of displacement and acceleration, but we have only one. We postpone this obsta-

TABLE 3.2 Adams–Stoermer Calculations

$T_n = 1.00$ sec	$u_0 = 1.00$ in	$\omega_n = 6.283$ rad/sec	
$\zeta = 0.05$	$\dot{u}_0 = 0.00$	$h = 0.125$ sec	
1	2	3	4
t (sec)	u (in)	$\dot{u}$ (in/sec)	$\ddot{u}$ (in/sec^2)
0.000	1.0000	0.000	−39.48
0.125	0.7145	−4.273	−25.52
0.250	0.0303	−6.592	2.95
0.375	−0.6080	−4.444	26.79
0.500	−0.8275	0.396	32.42
0.625	−0.5405	4.800	18.32
0.750	0.0328	6.210	−5.19
0.875	0.5249	4.090	−23.29
1.000	0.6531	0.048	−25.81

cle for later consideration, circumventing it for this illustration by the usually unavailable device of using the exact solution for the first time step, which the second line of the table shows. For time steps beyond the first, we enter the time in column 1, find the displacement and velocity from Eqs. (3.4.7) and (3.4.8) and enter them in columns 2 and 3, respectively, and finally use these in the differential equation of motion to get the acceleration in column 4, which completes the time step.

The solid triangles in Fig. 3.1 are the displacements computed by the Adams–Stoermer method. The deviation from the exact solution seems to be a bit greater than for the linear-acceleration method, and this time the numerical solution leads the exact solution.

The Adams–Stoermer process is simple indeed. Three straightforward calculations complete a time step, with error terms of the same order as for the linear-acceleration method, and without iteration. Alas, the benefits are not without cost. First, because the recurrence relations contain u_{n-1} and $\ddot{u}_{n-1}$, some other process must be employed for the first step. That is no particular bother for a hand calculation, but the computer program for a single step takes almost as many instructions as for N steps. Second, the use of past values of displacement and acceleration in the recurrence formulas restricts our freedom to change the length of the time step h during the solution. We may double the time step or increase it by any integer multiple by retaining a record of recent past displacements and accelerations and altering the instructions for the first occurrence of the greater time step; however, there is no way to reduce the time step or to adjust it by a noninteger multiple except by using an interpolation procedure for establishing the needed u_{n-1} and $\ddot{u}_{n-1}$.

The error terms dropped from Eqs. (3.4.3) and (3.4.6) all contain fourth and higher derivatives; thus, the method would be exact if the acceleration varied linearly over the double interval $t_{n-1} < t < t_{n+1}$.

The Adams–Stoermer method is especially attractive if velocity is absent from the differential equation of motion. In that case, the velocity-recurrence relation is not needed at all, and the velocity calculation may be omitted entirely.

3.5 LINEAR-ACCELERATION PROGRAM

Program 3.1, written in BASIC, solves the differential equation of motion for a single-degree-of-freedom system by the linear-acceleration method. It calls upon subroutines to provide the input data that defines the system properties and the time interval h, to compute accelerations, and to record the results.

Program 3.1: Linear-Acceleration Method
$$[\ddot{u} = f(u, \dot{u}, t)]$$

```
10   GOSUB 1000        ' Get data
20   GOSUB 2000        ' Start solution
30   WHILE T<TEND
40     GOSUB 5000      ' Advance one time step
50   WEND
60   END


1000                   ' Housekeeping instructions and data

2000                   ' Starter instructions

3000                   ' Acceleration subroutine

4000                   ' Output subroutine

5000                     ' Linear-acceleration subroutine
5010 UX = U + H*(V+H*A/3)
5020 VX = V + H*A/2
5030 T = T + H
5040 CHANGE = 1+EPSILON
5050 WHILE CHANGE>EPSILON
5060   AX = A
5070   U = UX + H*H*A/6
5080   V = VX + H*A/2
5090   GOSUB 3000    ' Find acceleration
5100   CHANGE = ABS(AX-A)
5110 WEND
5120 GOSUB 4000      ' Record results
5130 RETURN
```

Statements 10–60 control the program, calling the various subroutines as needed.

The subroutine at statement 1000 must establish the system and driving-force parameters, the time interval h, the ending time for the solution, and a convergence parameter EPSILON. Requiring convergence to zero is not feasible, for the iteration might alternate between two small values and never reach zero.

Subroutine 2000 sets the initial values of T, U, V, and A and prints headings and nonrecurring output. Subroutine 3000 computes the acceleration for the current values of T, U, and V, and subroutine 4000 records the output for the current time step.

Subroutine 5000–5130 is the main linear-acceleration subroutine. It first computes and saves the unchanging parts of the recurrence relations, advances time, and then iteratively completes the recurrence relations for displacement and velocity, computes the new acceleration, and compares it with the trial acceleration, continuing the iteration until the discrepancy is smaller than the convergence parameter EPSILON.

3.6 ADAMS–STOERMER PROGRAM

BASIC Program 3.2 solves the differential equation of motion for a single-degree-of-freedom system by the Adams–Stoermer method. Like the linear-acceleration program, it calls upon subroutines to provide the input data that defines the system properties, the initial conditions, and the time interval h, to compute accelerations, and to record the results. Additionally, it calls upon a subroutine to calculate the first step, which the Adams–Stoermer method is unable to do.

Program 3.2: Adams–Stoermer Method
$[\ddot{u} = f(u, \dot{u}, t)]$

```
10    GOSUB 1000        ' Get data
20    GOSUB 2000        ' Start solution
30    WHILE T<TEND
40      GOSUB 5000      ' Advance one time step
50    WEND
60    END


1000                    ' Housekeeping instructions and data

2000                    ' Starter instructions

3000                    ' Acceleration subroutine

4000                    ' Output subroutine

5000                    ' Adams-Stoermer subroutine
5010  T=T+H
5020  U=2*UN-UN1+H^2*AN
5030  V=V+.5*H*(3*AN-AN1)
5040  GOSUB 3000        ' Find acceleration
5050  GOSUB 4000        ' Record results
5060  UN1=UN
5070  UN=U
5080  AN1=AN
5090  AN=A
5100  RETURN
```

Again statements 10–60 control the program, calling the various subroutines as needed. Subroutines 1000–4000 are essentially the same as for Program 3.1 except that subroutine 2000 must generate the first time step, store the first two values of displacement as UN1 and UN, and store the first two values of acceleration as AN1 and AN. Subroutine 2000 could well require more instructions than any other subroutine.

Subroutine 5000–5100 is the main Adams–Stoermer subroutine. At entry, T, UN, V, and AN must be the values of time, displacement, velocity, and acceleration at the end of the last previous time step, respectively, and UN1 and AN1 the displacement and acceleration, respectively, one step earlier. The subroutine advances the solution one time step and updates all of these variables.

3.7 SPECIAL LINEAR-ACCELERATION METHOD

For the linear oscillator described by the differential equation

$$\ddot{u} + 2\zeta\omega\dot{u} + \omega^2 u = f(t)/m \tag{3.7.1}$$

we may alter the linear-acceleration method to circumvent the iteration. In Eq. (3.7.1), we designate the frequency as ω instead of the customary ω_n to avoid confusion with the time subscript. The differential equation evaluated at time t_{n+1} is

$$\ddot{u}_{n+1} = f(t_{n+1})/m - 2\zeta\omega\dot{u}_{n+1} - \omega^2 u_{n+1} \tag{3.7.2}$$

We get u_{n+1} and $\dot{u}_{n+1}$ from Eqs. (3.3.5) and (3.3.6), insert them into the right side of Eq. (3.7.2), and rearrange terms to get Eq. (3.7.3) that follows, which gives us the acceleration at the end of the interval in terms of the driving force at both ends of the interval and conditions at the beginning of the interval. The three steps, in the order given, complete a time step without iteration:

1. $\ddot{u}_{n+1} = \dfrac{[f(t_{n+1}) - f(t_n)]/m - \omega^2 h\dot{u}_n + [1 - \omega h(\zeta + \omega h/3)]\ddot{u}_n}{1 + \omega h(\zeta + \omega h/6)}$ (3.7.3)

2. $\dot{u}_{n+1} = \dot{u}_n + (h/2)(\ddot{u}_n + \ddot{u}_{n+1})$ (3.7.4)

3. $u_{n+1} = u_n + h\dot{u}_n + (h^2/6)(2\ddot{u}_n + \ddot{u}_{n+1})$ (3.7.5)

We emphasize that this is a **special** linear-acceleration method, valid only for the single-degree-of-freedom linear system described by Eq. (3.7.1).

3.8 PROGRAM FOR SPECIAL LINEAR-ACCELERATION METHOD

BASIC Program 3.3 solves the damped linear-oscillator problem described by Eq. (3.7.1), calling upon subroutines to define the system properties, the initial conditions, the driving force, and the time interval h, and to record the results.

Program 3.3: Special Linear-Acceleration Method
$$[\ddot{u} = f(t)/m - 2\zeta\omega_n\dot{u} - \omega_n^2 u]$$

```
10    GOSUB 1000         ' Get data
20    C1=-OMEGA^2*H
30    C2=1-OMEGA*H*(ZETA+OMEGA*H/3)
40    C3=1+OMEGA*H*(ZETA+OMEGA*H/6)
50    GOSUB 2000         ' Start solution
60    WHILE T<TEND
70      GOSUB 5000       ' Advance one time step
80    WEND
90    END

1000                     ' Housekeeping instructions and data

2000                     ' Starter instructions

3000                     ' Driving force subroutine

4000                     ' Output subroutine

5000                     ' Special lin-accel subroutine
5010  T = T+H
5020  OLDF = F
5030  OLDU = U
5040  OLDV = V
5050  OLDA = A
5060  GOSUB 3000         ' Get driving force
5070  A = ((F-OLDF)/M + C1*OLDV + C2*OLDA)/C3
5080  V = OLDV+H*(OLDA+A)/2
5090  U = OLDU+H*(OLDV+H*(2*OLDA+A)/6)
5100  GOSUB 4000         ' Record results
5110  RETURN
```

The control statements for this program include computation of the values of the coefficients C1, C2, and C3, which are used in the main special linear-acceleration subroutine, statement 5070. These coefficients depend upon the value of H as well as the system properties, so if H is changed during the solution, C1, C2, and C3 must be recalculated.

3.9 THE NEWMARK BETA METHOD

The Beta method might be considered a generalization of the linear-acceleration method. It employs a coefficient β set by the user. We choose a coefficient β and a time step h and then, having the values of u_n, $\dot{u}_n$, and $\ddot{u}_n$ at time t_n:

1. Estimate $\ddot{u}_{n+1}^*$, the acceleration at the end of the interval. The beginning acceleration $\ddot{u}_n$ is ordinarily the simplest first estimate of $\ddot{u}_{n+1}^*$.

2. Compute

$$u_{n+1} = u_n + h\dot{u}_n + h^2[(\tfrac{1}{2} - \beta)\ddot{u}_n + \beta\ddot{u}_{n+1}^*)] \qquad (3.9.1)$$

3. Compute

$$\dot{u}_{n+1} = \dot{u}_n + (h/2)(\ddot{u}_n + \ddot{u}_{n+1}^*) \qquad (3.3.6)$$

4. Compute

$$\ddot{u}_{n+1} = f(u_{n+1}, \dot{u}_{n+1}, t_{n+1}) \qquad (3.1.1)$$

5. (a) If $\ddot{u}_{n+1} \neq \ddot{u}_{n+1}^*$, take the computed $\ddot{u}_{n+1}$ as an improved estimate $\ddot{u}_{n+1}^*$ and return to step 2.
   (b) If $\ddot{u}_{n+1} = \ddot{u}_{n+1}^*$, the iteration has converged. Advance to the next time interval and return to step 1.

Again, if Eq. (3.1.1) does not contain the velocity term, it is more efficient to defer step 3 until the iteration has converged.

The coefficient β may be assigned any value in the range $0 \leq \beta \leq \tfrac{1}{2}$. With $\beta = \tfrac{1}{6}$, the process is identical to the linear-acceleration method and would be exact if the acceleration varied linearly across the time interval; $\beta = \tfrac{1}{4}$ would be exact if the velocity varied linearly, which would require that the mean acceleration prevail for the entire interval; $\beta = \tfrac{1}{8}$ would be exact if the acceleration retained its initial value for the first half of the interval and then jumped to its final value for the second half. Still other interpretations could be made for other values of β.

The best choice of β is an enigma. The Taylor series derivation might suggest that $\beta = \tfrac{1}{6}$ would give the most accurate results, but that is not necessarily so. For the case of free vibration of a damped linear oscillator, for example, $\beta = \tfrac{1}{12}$ gives near optimum accuracy. Newmark suggests that a value in the range $\tfrac{1}{6} \leq \beta \leq \tfrac{1}{4}$ is satisfactory from all points of view, including that of accuracy, and that $\beta = \tfrac{1}{6}$ assures stability of the method even when applied to multidegree-of-freedom systems.

3.10 PROGRAM FOR NEWMARK BETA METHOD

The Newmark Beta method program, Program 3.4, is nearly the same as Program 3.1. The only differences are the modifications of statements 5010 and 5070 and the inclusion of BETA among the input parameters.

Program 3.4: Newmark Beta Method
$$[\ddot{u} = f(u, \dot{u}, t)]$$

```
10    GOSUB 1000        ' Get data
20    GOSUB 2000        ' Start solution
30    WHILE T<TEND
40        GOSUB 5000    ' Advance one time step
50    WEND
60    END


1000                    ' Housekeeping instructions and data

2000                    ' Starter instructions

3000                    ' Acceleration subroutine

4000                    ' Output subroutine

5000                    ' Beta-method subroutine
5010  UX = U + H*(V+H*(1/2-BETA)*A)
5020  VX = V + H*A/2
5030  T = T + H
5040  CHANGE = 1+EPSILON
5050  WHILE CHANGE>EPSILON
5060      AX = A
5070      U = UX + H*H*BETA*A
5080      V = VX + H*A/2
5090      GOSUB 3000    ' Find acceleration
5100      CHANGE = ABS(AX-A)
5110  WEND
5120  GOSUB 4000        ' Record results
5130  RETURN
```

3.11 A RUNGE–KUTTA METHOD

There are several practical variants of the Runge–Kutta class of methods. As applied to the differential equations of motion, all of them involve making several trial calculations of acceleration using information within a time interval and then taking weighted averages of those accelerations to project the displacement and velocity to the end of the interval. The methods are single step and noniterative. Third- and fourth-order variants are available. We present here one fourth-order process particularly well adapted to the equations of motion. For the equation of motion

$$\ddot{u} = f(u, \dot{u}, t) \tag{3.1.1}$$

we make the calculations for one time interval as follows:

1. Compute

$$k_1 = hf(u_n, \dot{u}_n, t_n) \tag{3.11.1}$$

2. Compute

$$k_2 = hf(u_n + h\dot{u}_n/2 + hk_1/8, \; \dot{u}_n + k_1/2, \; t_n + h/2) \tag{3.11.2}$$

3. Compute

$$k_3 = hf(u_n + h\dot{u}_n/2 + hk_1/8, \; \dot{u}_n + k_2/2, \; t_n + h/2) \tag{3.11.3}$$

4. Compute

$$k_4 = hf(u_n + h\dot{u}_n + hk_3/2, \; \dot{u}_n + k_3, \; t_n + h) \tag{3.11.4}$$

5. Compute

$$u_{n+1} = u_n + h[\dot{u}_n + (k_1 + k_2 + k_3)/6] \tag{3.11.5}$$

6. Compute

$$\dot{u}_{n+1} = \dot{u}_n + [k_1 + 2(k_2 + k_3) + k_4]/6 \tag{3.11.6}$$

Thus, in steps 1 through 4, we calculate four trial changes in velocity, k_1, k_2, k_3, and k_4, each of which involves information already known. We then use weighted averages of these in steps 5 and 6 to project the displacement and velocity, respectively, to the end of the time interval. While the process is horrid for hand calculation, it is readily programmed for the computer and it is a well-behaved process that offers true fourth-order accuracy in both the displacement and velocity calculations.

For the general undamped system having a differential equation of motion

$$\ddot{u} = f(u, t) \tag{3.11.7}$$

the process may be simplified to

1. Compute

$$k_1 = hf(u_n, t_n) \tag{3.11.8}$$

2. Compute

$$k_2 = hf(u_n + h\dot{u}_n/2 + hk_1/8, t_n + h/2) \qquad (3.11.9)$$

3. Compute

$$k_3 = hf(u_n + h\dot{u}_n + hk_2/2, t_n + h) \qquad (3.11.10)$$

4. Compute

$$u_{n+1} = u_n + h[\dot{u}_n + (k_1 + 2k_2)/6] \qquad (3.11.11)$$

5. Compute

$$\dot{u}_{n+1} = \dot{u}_n + (k_1 + 4k_2 + k_3)/6 \qquad (3.11.12)$$

3.12 RUNGE–KUTTA PROGRAMS

Program 3.5 is a Runge–Kutta program for a damped system, and Program 3.6 shows the simpler Runge–Kutta program for an undamped system. Both call upon subroutines to provide the input data, compute accelerations, and record results.

Program 3.5: Runge–Kutta Method
$[\ddot{u} = f(u, \dot{u}, t)]$

```
10     GOSUB 1000        ' Get data
20     GOSUB 2000        ' Start solution
30     WHILE T<TEND
40        GOSUB 5000     ' Advance one time step
50     WEND
60     END


1000                     ' Housekeeping instructions and data

2000                     ' Starter instructions

3000                     ' Acceleration subroutine

4000                     ' Output subroutine

5000                     ' Runge-Kutta subroutine
5010   OLDU=U
5020   OLDV=V
5030   K1=H*A
5040   T=T+H/2
5050   U=OLDU+H*(OLDV/2+K1/8)
```

```
5060  V=OLDV+K1/2
5070  GOSUB 3000       ' Acceleration
5080  K2=H*A
5090  V=OLDV+K2/2
5100  GOSUB 3000       ' Acceleration
5110  K3=H*A
5120  T=T+H/2
5130  U=OLDU+H*(OLDV+K3/2)
5140  V=OLDV+K3
5150  GOSUB 3000       ' Acceleration
5160  K4=H*A
5170  U=OLDU+H*(OLDV+(K1+K2+K3)/6)
5180  V=OLDV+(K1+2*(K2+K3)+K4)/6
5190  GOSUB 3000       ' Acceleration
5200  GOSUB 4000       ' Record results
5210  RETURN
```

Subroutine 5000–5210 makes four projections in each time step, calling the acceleration subroutine four times, and uses weighted averages of them to get the end displacement and velocity.

Program 3.6: Runge–Kutta Instructions for Undamped System [$\ddot{u} = f(u, t)$]

```
10    GOSUB 1000       ' Get data
20    GOSUB 2000       ' Start solution
30    WHILE T<TEND
40      GOSUB 5000     ' Advance one time step
50    WEND
60    END


1000                   ' Housekeeping instructions and data

2000                   ' Starter instructions

3000                   ' Acceleration subroutine

4000                   ' Output subroutine

5000                   ' Runge-Kutta subroutine, undamped

5010  OLDU=U
5020  OLDV=V
5030  K1=H*A
5040  T=T+H/2
5050  U=OLDU+H*(OLDV/2+K1/8)
5060  GOSUB 3000       ' Acceleration
```

```
5070  K2=H*A
5080  T=T+H/2
5090  U=OLDU+H*(OLDV+K2/2)
5100  GOSUB 3000          ' Acceleration
5110  K3=H*A
5120  U=OLDU+H*(OLDV+(K1+2*K2)/6)
5130  V=OLDV+(K1+4*K2+K3)/6
5140  GOSUB 3000          ' Acceleration
5150  GOSUB 4000          ' Print
5160  RETURN
```

This Runge–Kutta subroutine 5000–5160 makes only three projections in each time step instead of the four projections of Program 3.5. The results are the same as those computed by the Program 3.5, for if that program is used for an undamped system, two of the four projections are identical.

3.13 A MILNE PREDICTOR–CORRECTOR METHOD

W. E. Milne developed fourth-order methods that are straightforward and noniterative like the Adams–Stoermer method, but which use additional past values to obtain a higher order of accuracy. For the system described by Eq. (3.1.1), we can derive Milne's recurrence relations from a Taylor series in the following manner.

Projecting velocity forward a distance $2h$ from time t_{n-1} in a Taylor series expansion gives us

$$\dot{u}_{n+1} = \dot{u}_{n-1} + (2h)\ddot{u}_{n-1} + \frac{(2h)^2}{2!}\dddot{u}_{n-1} + \frac{(2h)^3}{3!}u^{iv}_{n-1}$$
$$+ \frac{(2h)^4}{4!}u^{v}_{n-1} + \frac{(2h)^5}{5!}u^{vi}_{n-1} + \ldots$$

Similarly, projecting backwards a distance $2h$ from t_{n-1} yields

$$\dot{u}_{n-3} = \dot{u}_{n-1} - (2h)\ddot{u}_{n-1} + \frac{(2h)^2}{2!}\dddot{u}_{n-1} - \frac{(2h)^3}{3!}u^{iv}_{n-1}$$
$$+ \frac{(2h)^4}{4!}u^{v}_{n-1} - \frac{(2h)^5}{5!}u^{vi}_{n-1} + \ldots$$

We subtract the backward expansion from the forward expansion to get

$$\dot{u}_{n+1} - \dot{u}_{n-3} = 4h\ddot{u}_{n-1} + (8h^3/3)u^{iv}_{n-1} + (8h^5/15)u^{vi}_{n-1} + \ldots \qquad (3.13.1)$$

Projecting the acceleration forward and backward a distance h from time t_{n-1} in the same manner and adding the two projections leads to

$$\ddot{u}_n + \ddot{u}_{n-2} = 2\ddot{u}_{n-1} + h^2 u_{n-1}^{iv} + (h^4/12)u_{n-1}^{vi} + \ldots \qquad (3.13.2)$$

We find u_{n-1}^{iv} from Eq. (3.13.2), put it into Eq. (3.13.1), and rearrange terms to get

$$\dot{u}_{n+1} = \dot{u}_{n-3} + (4h/3)(2\ddot{u}_n - \ddot{u}_{n-1} + 2\ddot{u}_{n-2})$$
$$- (14/45)h^5 u_{n-1}^{vi} + \ldots \qquad (3.13.3)$$

Equation (3.13.3) expresses the velocity at the end of the time step in terms of quantities known at the beginning of the step.

Now we undertake a similar maneuver, projecting the displacement forward and backward a distance h from time t_n. The forward projection gives us

$$u_{n+1} = u_n + h\dot{u}_n + (h^2/2)\ddot{u}_n + (h^3/3!)\dddot{u}_n$$
$$+ (h^4/4!)u_n^{iv} + (h^5/5!)u_n^{v} + \ldots$$

and the backward projection

$$u_{n-1} = u_n - h\dot{u}_n + (h^2/2)\ddot{u}_n - (h^3/3!)\dddot{u}_n$$
$$+ (h^4/4!)u_n^{iv} - (h^5/5!)u_n^{v} + \ldots$$

We subtract the backward projection from the forward projection to get

$$u_{n+1} - u_{n-1} = 2h\dot{u}_n + (2h^3/3!)\dddot{u}_n + (2h^5/5!)u_n^{v} + \ldots \qquad (3.13.4)$$

In a similar vein, projecting the velocity forward and backward a distance h from t_n yields the term-by-term derivatives of the two foregoing projections, which we add to get

$$\dot{u}_{n+1} + \dot{u}_{n-1} = 2\dot{u}_n + h^2\dddot{u}_n + (2h^4/4!)u_n^{v} + \ldots \qquad (3.13.5)$$

We substitute $\dddot{u}_n$ from Eq. (3.13.5) into Eq. (3.13.4) to get

$$u_{n+1} = u_{n-1} + (h/3)(\dot{u}_{n-1} + 4\dot{u}_n + \dot{u}_{n+1}) - (1/90)h^5 u_n^{v} + \ldots \qquad (3.13.6)$$

which we differentiate term by term to obtain

$$\dot{u}_{n+1} = \dot{u}_{n-1} + (h/3)(\ddot{u}_{n-1} + 4\ddot{u}_n + \ddot{u}_{n+1})$$
$$- (1/90)h^5 u_n^{vi} + \ldots \qquad (3.13.7)$$

Equations (3.13.3), (3.13.6), and (3.13.7) all have error terms of order h^5, but Eq. (3.13.3) has a much larger coefficient of h^5. On the other hand, Eqs. (3.13.6) and (3.13.7) contain quantities not known at the beginning of the interval.

Milne's predictor–corrector method first predicts the velocity and displacement at the end of the interval from Eqs. (3.13.3) and (3.13.6), then uses these predicted values in Eq. (3.1.1) to predict the end acceleration, next corrects the end velocity and displacement with Eqs. (3.13.7) and (3.13.6), and finally invokes Eq. (3.1.1) again to correct the end acceleration. One time step of the process is

PREDICT

1. $\qquad \dot{u}^*_{n+1} = \dot{u}_{n-3} + (4h/3)\,(2\ddot{u}_n - \ddot{u}_{n-1} + 2\ddot{u}_{n-2})$ $\qquad$ (3.13.8)

2. $\qquad u^*_{n+1} = u_{n-1} + (h/3)\,(\dot{u}_{n-1} + 4\dot{u}_n + \dot{u}^*_{n+1})$ $\qquad$ (3.13.9)

3. $\qquad \ddot{u}^*_{n+1} = f(u^*_{n+1}, \dot{u}^*_{n+1}, t_{n+1})$ $\qquad$ (3.1.1)

CORRECT

4. $\qquad \dot{u}_{n+1} = \dot{u}_{n-1} + (h/3)\,(\ddot{u}_{n-1} + 4\ddot{u}_n + \ddot{u}^*_{n+1})$ $\qquad$ (3.13.10)

5. $\qquad u_{n+1} = u_{n-1} + (h/3)\,(\dot{u}_{n-1} + 4\dot{u}_n + \dot{u}_{n+1})$ $\qquad$ (3.13.11)

6. $\qquad \ddot{u}_{n+1} = f(u_{n+1}, \dot{u}_{n+1}, t_{n+1})$ $\qquad$ (3.1.1)

The Milne process is a true fourth-order process, for the error terms in Eqs. (3.13.3), (3.13.6), and (3.13.7) are all of order h^5. This differs from the linear-acceleration and Adams–Stoermer methods, in which the velocity equations had error terms of lower order than the displacement equations.

The method is fast and accurate. There is no iteration, for all of the quantities involved in each computational step are known when the computation is made. This differs from the linear-acceleration and Newmark Beta methods, which employ an estimated end acceleration and then iterate until convergence is reached. However, Milne uses information from three past steps, and therefore requires that the first three steps of the solution be accomplished some other way. For all practical purposes, the time step h must remain unchanged throughout the solution.

3.14 PREDICTOR–CORRECTOR PROGRAM

Program 3.7 solves Eq. (3.1.1) by the Milne Predictor–Corrector method, calling on subroutines to provide the input data and time interval and to calculate accelerations and record results. It also calls upon an unspecified subroutine to compute three time steps.

Program 3.7: Milne Predictor–Corrector Method
$[\ddot{u} = f(u, \dot{u}, t)]$

```
10    GOSUB 1000         ' Get data
20    GOSUB 2000         ' Start solution
30    WHILE T<TEND
40       GOSUB 5000      ' Advance one time step
50    WEND
60    END


1000                     ' Housekeeping instructions and data

2000                     ' Starter instructions

3000                     ' Acceleration subroutine

4000                     ' Output subroutine

5000                     ' Milne Predictor-Corrector routine
5010  V=VN(3)+4*H/3*(2*(AN(2)+AN(0))-AN(1))
5020  U=UN(1)+H/3*(VN(1)+4*VN(0)+V)
5030  T=T+H
5040  GOSUB 3000         ' Find acceleration
5050  V=VN(1)+H/3*(AN(1)+4*AN(0)+A)
5060  U=UN(1)+H/3*(VN(1)+4*VN(0)+V)
5070  GOSUB 3000         ' Find acceleration
5080  GOSUB 4000         ' Record results
5090  FOR I=3 TO 1 STEP -1
5100     AN(I)=AN(I-1)
5110     UN(I)=UN(I-1)
5120     VN(I)=VN(I-1)
5130  NEXT I
5140  AN(0)=A
5150  VN(0)=V
5160  UN(0)=U
5170  RETURN
```

Here control statements 10–60 and subroutines 1000–4000 are essentially the same as for the Adams–Stoermer program, Program 3.2, except that subroutine 2000 must generate the first three time steps and store the displacements, velocities, and accelerations as subscripted variables: UN(3), . . . , UN(0); VN(3), . . . , VN(0); and AN(3), . . . , AN(0). Again, subroutine 2000 could well require more instructions than any other subroutine.

Subroutine 5000–5100 is the main Predictor–Corrector subroutine. It advances the solution one time step and updates variables UN(3), . . . , UN(0), VN(3), . . . , VN(0), and AN(3), . . . , AN(0).

3.15 A MILNE PREDICTOR–CORRECTOR METHOD FOR UNDAMPED SYSTEMS

If the system is undamped, so that its differential equation of motion does not contain velocity, we cannot simply delete the velocity-recurrence relation from the Milne process the way we could with the Adams–Stoermer process. However, the following Milne method for undamped systems, also a fourth-order process, uses recurrence relations that eliminate the need for computing the velocity. The procedure requires having the values of u for three preceding steps and $\ddot{u}$ for two. With that information, the procedure for one time step is

PREDICT

1. $u_{n+1}^* = u_n + u_{n-2} - u_{n-3} + (h^2/4)(5\ddot{u}_n + 2\ddot{u}_{n-1} + 5\ddot{u}_{n-2})$ (3.15.1)

2. $\ddot{u}_{n+1}^* = f(u_{n+1}^*, t_{n+1})$ (3.11.7)

CORRECT

3. $u_{n+1} = 2u_n - u_{n-1} + (h^2/12)(\ddot{u}_{n+1}^* + 10\ddot{u}_n + \ddot{u}_{n-1})$ (3.15.2)

4. $\ddot{u}_{n+1} = f(u_{n+1}, t_{n+1})$ (3.11.7)

3.16 STARTING MULTIPLE-STEP METHODS

Both the Adams–Stoermer and Milne Predictor–Corrector methods require that the initial steps be calculated some other way. Because any error generated in the initial steps will affect the entire solution, those steps must be computed as accurately as the rest of the solution. The linear-acceleration method has the same order of accuracy as Adams–Stoermer in both the displacement and velocity projections, and thus could be used as its starter. Similarly, we could use the Runge–Kutta method to start the Milne Predictor–Corrector method, for they have the same order of accuracy. Alternatively, we could use the linear-acceleration method with a smaller time step.

A Taylor series may provide a suitable alternative starting method in some cases. The series expansions about time $t = 0$ for $u(\tau)$ and $\dot{u}(\tau)$ are

$$u_n = u_0 + \tau\dot{u}_0 + (\tau^2/2)\ddot{u}_0 + (\tau^3/3!)\dddot{u}_0 + (\tau^4/4!)u_0^{iv} + \ldots$$
$$\dot{u}_n = \dot{u}_0 + \tau\ddot{u}_0 + (\tau^2/2)\dddot{u}_0 + (\tau^3/3!)u_0^{iv} + (\tau^4/4!)u_0^{v} + \ldots$$

If successive derivatives can be calculated, the two series may be carried out until the terms become too small to alter the retained values.

Consider the example we have been using, a damped linear oscillator in free vibration. The differential equation is

$$\ddot{u} = -2\zeta\omega\dot{u} - \omega^2 u$$

The initial conditions give us u_0 and $\dot{u}_0$. Higher derivatives can be computed as

$$u^{(n)} = -2\zeta\omega u^{(n-1)} - \omega^2 u^{(n-2)}$$

The calculation is readily programmed for computer generation of the successive terms of a Taylor series. Program 3.8 computes the first three time steps, continuing the series until the next term becomes too small to affect either the displacement or the velocity. It records the displacements $u_n, \ldots, u_{n-3}$ as UN(0), ..., UN(3), the velocities as VN(0), ..., VN(3), and the accelerations as AN(0), ..., AN(3), as we used them in the Milne program, Program 3.7.

Program 3.8: Taylor Series Starter for Free-Vibration Problem

```
2000                        ' Starter  instructions
2010 FOR I=0 TO 3
2020    T=I*H
2030    U=U0
2040    V=V0
2050    GOSUB 3000          ' Find  acceleration
2060    D0=U
2070    D1=V
2080    D2=A
2090    COEFF=T
2100    TERM=1
2110    OLDU=1+U
2120    WHILE OLDU<>U OR OLDV<>V
2130       D0=D1
2140       D1=D2
2150       D2=OMEGA*(2*ZETA*D1+OMEGA*D0)
2160       OLDU=U
2170       OLDV=V
2180       U=U+COEFF*D0
2190       V=V+COEFF*D1
2200       TERM=TERM+1
2210       COEFF=COEFF*T/TERM
2220    WEND
2230    GOSUB 3000          ' Find  acceleration
2240    GOSUB 4000          ' Record  results
2250    UN(3-I)=U
```

```
2260    VN(3-I)=V
2270    AN(3-I)=A
2280 NEXT I
2290 RETURN
```

For even this simple problem, the starter instructions outnumber the instructions in the main integration subroutine. A driving force would complicate the process, for the force and its derivatives would have to be taken into account in computing successive terms of the series. If the driving force for the first few time steps could be approximated by a polynomial, then the complication would be minor because only the first few derivatives would be nonzero. In any case, a Taylor series starter must be written for the specific problem. There is no way to make it a general-purpose subroutine.

3.17 COMPUTATIONAL ERROR

Error is generated in solving any differential equation numerically. We will not go into error analysis, which is in itself a demanding mathematical discipline. Rather, we will examine the solutions of two specific problems by some of the numerical methods presented in this chapter and compare them with the exact analytical solutions to get a feel for the size of numerical error, and then consider an unsophisticated but useful way of managing error.

For the first comparison, we take a damped linear oscillator with 5 percent of critical damping, in free vibration with zero initial velocity and unit initial displacement. This is the vibration for which Fig. 3.1 compared two numerical solutions with the exact solution for the first few cycles. Those numerical solutions were computed using a time step of $h = T_n/8$. Here we compute the first ten cycles of oscillation for the same problem by several different methods, using time steps ranging from $T_n/8$ to $T_n/96$. Figure 3.2 shows the extreme error for each method, plotted against mesh size in a log–log plot. We define error to be the maximum difference between the numerically computed displacement and the exact displacement, divided by the maximum exact displacement, that is,

$$|u_{\text{computed}} - u_{\text{exact}}|/u_{\text{maximum}}$$

The linear-acceleration and Adams–Stoermer methods produce nearly the same error, and that error decreases approximately as h^2. Because they are nominally third-order methods one might expect the error to decrease as h^3, but the results are otherwise. Recall that for both methods, the displacement-recurrence relation had h^3 accuracy and the velocity-recurrence relation, h^2. Runge–Kutta has the smallest error, but the Milne Predictor–Corrector

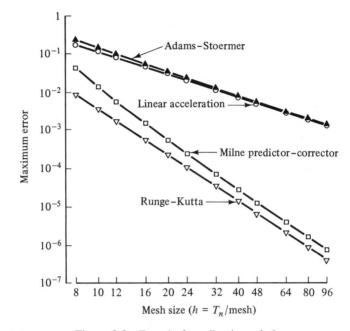

Figure 3.2 Error in free-vibration solution.

method is nearly as accurate, and for both Runge–Kutta and Milne, the error decreases approximately as h^4.

For a second comparison, we take an undamped spring–mass system having a period $T_n = 0.4$ sec and a mass $m = 100$ lbm and subject it to an initial peak triangular force pulse starting at $P_0 = 50$ lb and reducing linearly to zero at time $t_1 = 0.2$ sec. The initial displacement and velocity are zero. We carry out the solution to $t = 5$ sec, which is 12.5 times the undamped period. We use an undamped system so that we can employ all of the methods, including the second Milne Predictor–Corrector method, which cannot be applied to a damped system. Again we repeat the solution for time steps h ranging from $T_n/8$ to $T_n/96$.

Figure 3.3 shows the maximum error for each numerical method, again plotted as maximum error vs. mesh size on a log–log plot.

The linear-acceleration and Adams–Stoermer methods again give nearly the same accuracy, with the maximum error decreasing approximately as h^2. Runge–Kutta gives far the best accuracy, and its error decreases very nearly as h^4.

The Milne Predictor–Corrector methods, although they are both true fourth-order methods, produce a maximum error that varies approximately at $h^{2.5}$. The reason lies in the nature of the driving force, which ends abruptly at time t_1. The multistep method in effect treats the acceleration as though it varied smoothly over the nearby intervals, whereas the actual ac-

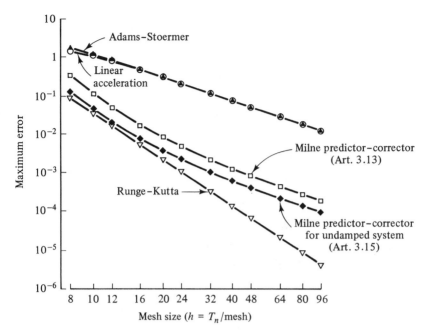

Figure 3.3   Error in forced-vibration solution.

celeration curve has a kink in it at the end of the pulse. This discrepancy injects an extra bit of error into several steps, and that error propagates through the solution.

Clearly, it is usually impossible to evaluate error by comparing the numerical solution with an exact solution. Also, it is usually impractical to employ rigorous error analysis to establish useful error bounds. The characteristics of the problem itself may decree a maximum time step; for example, the time distance between successive points on an earthquake accelerogram might dictate a maximum time step for computing the response of a structure to an earthquake.

One useful, albeit pedestrian, technique is to solve the problem with an arbitrary time step that appears reasonable, perhaps erring on the long side, then cut the time step in half, repeat the solution, and compare the results, continuing the process until two successive solutions are in adequate agreement.

3.18 NONLINEAR EXAMPLE

We now use numerical methods to solve a problem which, unlike earlier examples, is not amenable to analytical solution. The problem is to find the maximum displacement of a nonlinear yielding single-degree-of-freedom system subjected to an impulsive force. We represent the driving force as an initial peak triangular pulse, and we use a Ramberg–Osgood function to characterize the yield behavior. Figure 3.4 shows the system, driving force, and restoring force.

The form of the Ramberg–Osgood force-displacement relation is

$$Q/Q_y + \alpha(Q/Q_y)^N = u/u_y$$

where

Q = restoring force

Q_y = yield force

u = displacement

u_y = yield displacement

α = empirical coefficient, often taken to be about 0.1

N = odd integer exponent, often taken to be 7 or 9

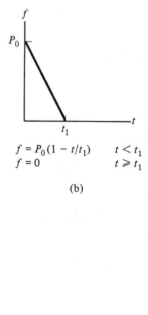

$$f = P_0(1 - t/t_1) \qquad t < t_1$$
$$f = 0 \qquad t \geqslant t_1$$

(b)

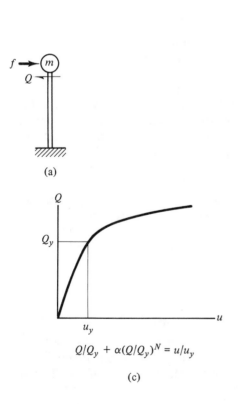

(a)

$$Q/Q_y + \alpha(Q/Q_y)^N = u/u_y$$

(c)

Figure 3.4 Example: (a) system, (b) driving force, and (c) restoring force.

At small displacements, the behavior is essentially elastic, with a stiffness Q_y/u_y. As the displacement approaches the yield displacement, the stiffness decreases, and at displacements beyond yield, the force-displacement relation approaches plastic behavior. The coefficient α determines the yield offset of the force-displacement curve from the initial elastic relation, and the exponent N determines how sharply the behavior changes from elastic to something approaching plastic behavior. N is taken to be an odd integer in order to preserve an odd functional relation between force and displacement. For very large N, the force-displacement relation approaches the ideal elastic–plastic relation.

The Ramberg–Osgood formulation could be redefined to make $\alpha = 1$ and to allow even integer or noninteger values of N, but we will adhere to the formulation given.

While the Ramberg–Osgood relation would make analytical solution a formidable task, it imposes no great burden on a numerical solution. However, it does present an obstacle in the implicit relation between Q and u. In each time step, the numerical procedure will give us a value of u and we will need the corresponding value of Q in order to compute the acceleration. We cannot manipulate the relation to obtain Q as an explicit function of u. Instead, we will have to solve for Q numerically in each time step. An efficient technique is Newton's method, which will also be useful for other purposes when we come to multidegree systems in Chapter 5.

Let

$$r = Q/Q_y, \text{ a dimensionless restoring force}$$

$$\xi = u/u_y, \text{ a dimensionless displacement}$$

The Ramberg–Osgood relation then becomes

$$r + \alpha r^N - \xi = 0$$

For any given value of ξ, we need to find the value of r that will make this function of r zero. Figure 3.5 shows the function in the vicinity of the root.

To apply Newton's method, we start with an approximate root, say r_0, evaluate the function and its derivative at $r = r_0$, find where the derivative crosses the zero axis, and take that value of r as an improved approximation r_1, as indicated in Fig. 3.5. We repeat the process until it converges to the root. The general Newton iteration is

$$r_{i+1} = r_i - f(r_i)/f'(r_i)$$

which for this particular example is

$$r_{i+1} = r_i - (r_i + \alpha r_i^N - \xi)/(1 + N\alpha r_i^{N-1})$$

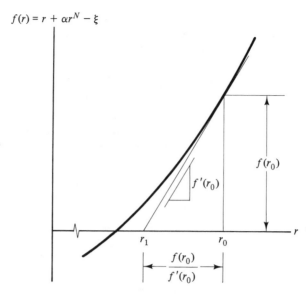

$$f(r) = r + \alpha r^N - \xi$$

Figure 3.5 Newton's method.

We can take the most recent computed value of r to be the initial approximate root r_0.

In this case, the derivative is easy to formulate and a study of Fig. 3.5 convinces us that the iteration will converge to the root. Alas, this is not always the case with Newton's method.

We want to find the maximum displacement, and when the velocity turns negative, we will know that a maximum was reached in the last time step. We could simply take the largest computed value of displacement to approximate the maximum. For a more accurate solution, we will use the final values of velocity and acceleration to estimate when during the last time step the velocity reached zero, compute a negative value of h to reverse the solution to that time, iterating if necessary, and thus get the maximum displacement within the time step. Single-step numerical methods allow the use of a negative time step to accomplish this. Multistep methods would not, and we would have to use an interpolation procedure to compute a maximum within the time step.

BASIC Program 3.9 solves the problem, accepting input data from the keyboard and writing output on the screen. The program uses the Runge–Kutta subroutine of Program 3.6 to perform the integration. It reads the input data and sets the time step, and uses subroutines to compute the driving force, to solve the Ramberg–Osgood relation, and to compute the acceleration. When the velocity turns negative, the program uses the final values of velocity and acceleration to estimate when during the final interval the velocity became zero, and backs up the solution to that time to get the maximum. Table 3.3 shows the output for one solution, which consists of the input data, the maximum displacement, and the time of the maximum displacement.

Either the linear-acceleration method or the Newmark Beta method could be used instead of Runge–Kutta simply by replacing the subroutine in instructions 5000–5150 and providing EPSILON and BETA among the input data.

Program 3.9: Response of Ramberg–Osgood Yielding System

```
10   GOSUB 1000              ' Get data
20   GOSUB 2000              ' Start solution
30   GOSUB 5000              ' Advance one time step
40   IF V>0 GOTO 30          ' Repeat as long as V>0
50   H=-V/A                  ' Go back to maximum displ
60   GOSUB 5000
70   IF ABS(V)>.001 GOTO 50  ' Repeat if maximum missed
80   GOSUB 4000              ' Record results
90   END


1000                   ' Housekeeping instructions and data
1010 CLS
1020 PI=3.141593
1030 GRAV=386.088
1040 INPUT "Wt, Tn:   ", WT, TN
1050 OMEGA=2*PI/TN
1060 M=WT/GRAV
1070 K=M*OMEGA^2
1080 INPUT "Qy, Alpha, N:  ", QY, ALPHA, N
1090 UY=QY/K
1100 INPUT "P0, T1:   ", P0, T1
1110 INPUT "Mesh size:  ", MESH
1120 H = TN/MESH
1130 RETURN


2000                   ' Starter instructions
2010 T = 0
2020 U = 0
2030 V = 0
2040 R = 0
2050 GOSUB 3000
2060 RETURN


3000                    ' Acceleration subroutine
3010 IF T<T1 THEN F=P0*(1-T/T1) ELSE F=0
```

```
3020 XI=U/UY
3030 DR=1
3040 WHILE ABS(DR)>.000001
3050   DR=(R+ALPHA*R^N-XI)/(1+N*ALPHA*R^(N-1))
3060   R=R-DR
3070 WEND
3080 Q=QY*R
3090 A=(F-Q)/M
3100 RETURN


4000                       ' Output subroutine
4010 CLS
4020 PRINT "System properties:"
4030 PRINT "Weight = "; WT
4040 PRINT "    Tn = "; TN
4050 PRINT " Omega = "; OMEGA
4060 PRINT "    Qy = "; QY
4070 PRINT " Alpha = "; ALPHA
4080 PRINT "     N = "; N
4090 PRINT "    Uy = "; UY
4100 PRINT: PRINT "Driving force:"
4110 PRINT "    P0 = "; P0
4120 PRINT "    T1 = "; T1
4130 PRINT: PRINT "  Mesh = "; MESH
4140 PRINT: PRINT "Maximum displacement = "; U;
4150 PRINT " at  t = "; T
4160 PRINT
4170 RETURN


5000                       ' Runge-Kutta subroutine, undamped
5010 OLDU=U
5020 OLDV=V
5030 K1=H*A
5040 T=T+H/2
5050 U=OLDU+H*(OLDV/2+K1/8)
5060 GOSUB 3000        ' Acceleration
5070 K2=H*A
5080 T=T+H/2
5090 U=OLDU+H*(OLDV+K2/2)
5100 GOSUB 3000        ' Acceleration
5110 K3=H*A
5120 U=OLDU+H*(OLDV+(K1+2*K2)/6)
5130 V=OLDV+(K1+4*K2+K3)/6
5140 GOSUB 3000        ' Acceleration
5150 RETURN
```

Table 3.3 Output of Program 3.9

```
System properties:
Weight =   100
    Tn =   .5
 Omega =   12.56637
    Qy =   75
 Alpha =   .1
     N =   9
    Uy =   1.833698

Driving force:
    P0 =   125
    T1 =   .4

 Mesh =   20

Maximum displacement =   6.693556   at   t =   .3148543
```

PROBLEMS

3.1 Write a program to solve Problem 1.7 numerically, and execute it for the data of Problem 1.6.

3.2 A mass m (considered to be a point mass) slides on a frictionless horizontal plane, as shown below. It is restrained by two weightless linear

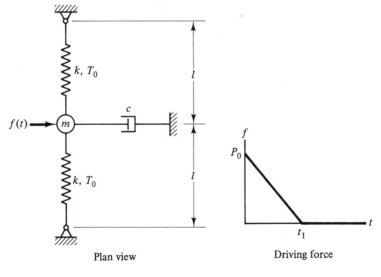

Plan view   Driving force

springs of equal length l anchored a distance $2l$ apart, and is damped by a weightless linear damper c. In the equilibrium position, the tension in the springs is T_0. The system is driven by an initial peak triangular pulse:

(a) Write a program to find the first maximum and first minimum displacements of the system.

(b) Find the first maximum and first minimum displacements for the following system properties:

$$m = 100 \text{ lbm } (m \text{ weighs } 100 \text{ lb})$$
$$c = 4 \text{ lb/(ft/sec)}$$
$$k = 600 \text{ lb/ft}$$
$$l = 3 \text{ ft}$$
$$T_0 = 40 \text{ lb}$$
$$P_0 = 60 \text{ lb}$$
$$t_1 = 0.5 \text{ sec}$$

3.3 A single-degree-of-freedom undamped elastic-plastic system, shown in the figure is driven by an impulsive force described by the equation

$$f(t) = P_0(1 - t/t_1)e^{-\alpha t/t_1}$$

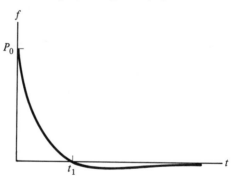

Driving force

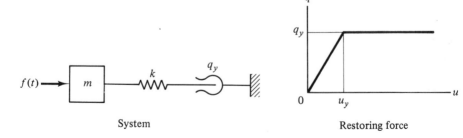

System

Restoring force

In the first surge of motion in the positive direction, the restoring force is

$$q = q_y(u/u_y) \qquad u < u_y$$
$$q = q_y \qquad u \geq u_y$$

(a) Write a program to find the extreme displacement of the system.
(b) Find the extreme displacement for the following system and driving-force properties:

$$m = 20{,}000 \text{ lbm } (m \text{ weighs } 20{,}000 \text{ lb})$$
$$q_y = 10{,}000 \text{ lb}$$
$$u_y = 0.75 \text{ in}$$
$$P_0 = 40{,}000 \text{ lb}$$
$$\alpha = 1.5$$
$$t_1 = 0.20 \text{ sec}$$

CHAPTER FOUR

Response Spectra

4.1 THE SPECTRUM CONCEPT

The response spectrum is an extremely useful concept in earthquake-engineering research and in applying the seismological knowledge of strong earthquakes to the design of structures and the development of building codes. Hugo Benioff initiated the application of spectrum concepts to seismology about 1934, and M. A. Biot extended the application to earthquake engineering in 1941. To make his analyses, Biot built a spectrum analyzer, an ingenious special-purpose mechanical analog computer, well before modern electronic analog and digital computers arrived on the scene.

The notion of a response spectrum is simple. The maximum response of a linear single-degree-of-freedom system to any given component of earthquake motion depends only on the natural frequency or period of the system and its fraction of critical damping. Response may be defined as displacement, velocity, or acceleration, either relative to the base or total. Any two systems having the same values of frequency and damping will have the same maximum response, even though one system may be more massive than the other. If we calculate the maximum response for a range of values of frequency and damping and portray the results graphically, we then have a spectrum chart that shows the maximum response of all possible linear SDF systems to that particular component of earthquake motion.

4.2 MATHEMATICAL RELATIONS

The equation for the response of a damped linear oscillator to an arbitrary base translation, with zero initial conditions, is given by Duhamel's integral:

$$u(t) = -(1/\omega_d) \int_0^t \ddot{u}_g(\tau) e^{-\zeta\omega_n(t-\tau)} \sin \omega_d(t - \tau) \, d\tau \qquad (4.2.1)$$

Using theorems from calculus, we may differentiate the integral in Eq. (4.2.1) under the integral sign. For more compact notation, let

$$S(t) = \int_0^t \ddot{u}_g(\tau) e^{-\zeta\omega_n(t-\tau)} \sin \omega_d(t - \tau) \, d\tau \qquad (4.2.2)$$

and

$$C(t) = \int_0^t \ddot{u}_g(\tau) e^{-\zeta\omega_n(t-\tau)} \cos \omega_d(t - \tau) \, d\tau \qquad (4.2.3)$$

Differentiating the integrals in Eqs. (4.2.2) and (4.2.3) under the integral signs, we get

$$\dot{S} = -\zeta\omega_n S + \omega_d C \qquad (4.2.4)$$

and

$$\dot{C} = -\zeta\omega_n C - \omega_d S + \ddot{u}_g(t) \qquad (4.2.5)$$

We incorporate these into Eq. (4.2.1) to get

$$u(t) = -(1/\omega_d)S(t) \qquad (4.2.6)$$

$$\dot{u}(t) = [\zeta/(1 - \zeta^2)^{1/2}]S(t) - C(t) \qquad (4.2.7)$$

$$\ddot{u}(t) + \ddot{u}_g(t) = [\omega_n(1 - 2\zeta^2)/(1 - \zeta^2)^{1/2}]S(t) + 2\zeta\omega_n C(t) \qquad (4.2.8)$$

Suitable manipulations with the trigonometric identities reduce Eqs. (4.2.6) to (4.2.8) to

$$u(t) = -(1/\omega_n) \left\{ [1/(1 - \zeta^2)^{1/2}] \int_0^t \ddot{u}_g(\tau) e^{-\zeta \omega_n (t - \tau)} \sin \omega_d(t - \tau)\, d\tau \right\}$$

(4.2.9)

$$\dot{u}(t) = \left\{ [1/(1 - \zeta^2)^{1/2}] \int_0^t \ddot{u}_g(\tau) e^{-\zeta \omega_n (t - \tau)} \sin [\omega_d(t - \tau) - \alpha]\, d\tau \right\}$$

(4.2.10)

where $\alpha = \tan^{-1} [(1 - \zeta^2)^{1/2}/\zeta]$.

$$\ddot{u} + \ddot{u}_g = \omega_n \left\{ [1/(1 - \zeta^2)^{1/2}] \int_0^t \ddot{u}_g(\tau) e^{-\zeta \omega_n (t - \tau)} \sin [\omega_d(t - \tau) - \beta]\, d\tau \right\}$$

(4.2.11)

where $\beta = \tan^{-1} [2\zeta(1 - \zeta^2)^{1/2}/(1 - 2\zeta^2)]$.

While Eqs. (4.2.9) to (4.2.11) may look very complicated, the significant feature is that they are all identical except for the initial multiplying factor and for a phase shift in the argument of the sine term in the integrand. The differences in the values of the three integrals at any time t is induced solely by the phase shift in the sine function. The phase shift in the acceleration equation, β in Eq. (4.2.11), is small, zero for $\zeta = 0$ and 11.5° for $\zeta = 0.10$. For the velocity equation, the phase shift, α in Eq. (4.2.10), is large, 90° for $\zeta = 0$ and 84° for $\zeta = 0.10$.

For engineering purposes, we are not especially concerned with the time variation of the response parameters. Rather, it is their extreme values that convey the crucial information, for they are related to the maximum forces, maximum displacements, and maximum deformations that structures must be able to endure. Analyses of many accelerograms of strong earthquakes have shown that the maximum magnitudes of the integrals in Eqs. (4.2.9) and (4.2.11) are nearly the same, that is, the small phase shift in the sine function does not have a pronounced influence on the extreme value of the integral. Indeed, even the large phase shift in Eq. (4.2.10) does not greatly alter the maximum value of the integral except for long-period systems.

Several different definitions of response spectra have been employed, but the following have come into fairly uniform use:

$$SD = |u|_{max} \qquad = \text{displacement response spectrum}$$

$$SV = |\dot{u}|_{max} \qquad = \text{velocity response spectrum}$$

$$SA = |\ddot{u} + \ddot{u}_g|_{max} = \text{acceleration response spectrum}$$

$$PSV = \omega_n \, SD \quad = \text{pseudo-velocity response spectrum}$$
$$PSA = \omega_n^2 \, SD \quad = \text{pseudo-acceleration response spectrum}$$

PSV and PSA are given the prefix "pseudo-" because they are not truly the extreme values of velocity and acceleration, although they have the correct dimensions. Usually, for strong-motion earthquakes, PSA and SA are nearly identical.

Table 4.1 shows the relations between these various response spectra and the extreme values of the corresponding measures of response.

TABLE 4.1. Response Spectrum Relations

Relative displacement $\|u\|_{max}$	$= SD$	$\simeq SV/\omega_n$	$\simeq SA/\omega_n^2$ *	$= PSV/\omega_n$	$= PSA/\omega_n^2$
Relative velocity $\|\dot{u}\|_{max}$	$\simeq \omega_n \, SD$	$= SV$	$\simeq SA/\omega_n$	$\simeq PSV$	$\simeq PSA/\omega_n$
Total acceleration $\|\ddot{u} + \ddot{u}_g\|_{max}$	$\simeq \omega_n^2 \, SD$ *	$\simeq \omega_n \, SV$	$= SA$	$\simeq \omega_n \, PSV$	$\simeq PSA$ *

*If $\zeta = 0$, these relations are exact.

4.3 SPECTRUM GRAPHS

It is convenient to show the spectrum as a family of curves, one for each of several different damping values, each portraying the maximum value of the response as a function of the natural frequency or period of the oscillator. Engineers seem to prefer to use natural period rather than natural frequency, perhaps because the period of vibration is a more familiar concept and one that has a stronger appeal to the intuition.

Figure 4.1 shows plots of the displacement, velocity, and acceleration spectra for one particular component of earthquake motion, the N11°W component of the earthquake recorded at Eureka, California, on December 21, 1954, for one damping value, $\zeta = 0.05$.

Figure 4.2(a) shows the velocity and pseudo-velocity response spectra, SV and PSV, respectively, and Fig. 4.2(b) shows SA and PSA. SA and PSA are so nearly the same in Fig. 4.2(b) that the PSA points appear to fall on the SA line over the entire period range. SV and PSV are close in the short-period range, but differ substantially for long-period systems. This is

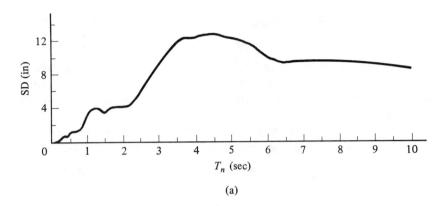

(a)

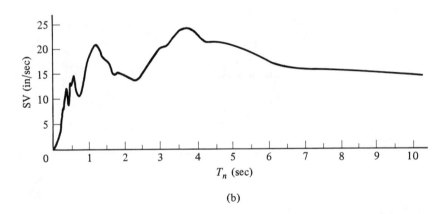

(b)

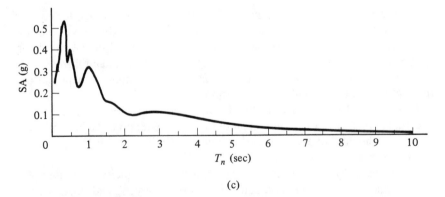

(c)

Figure 4.1   Response spectra ($\zeta = 0.05$) for N11°W component of Eureka, California, earthquake of December 21, 1954. (a) Displacement spectrum SD. (b) Velocity spectrum SV. (c) Acceleration spectrum SA.

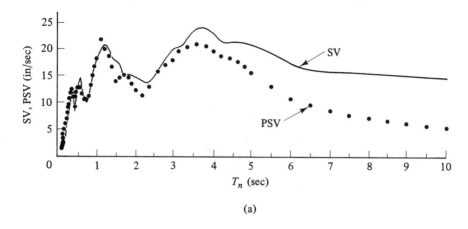

(a)

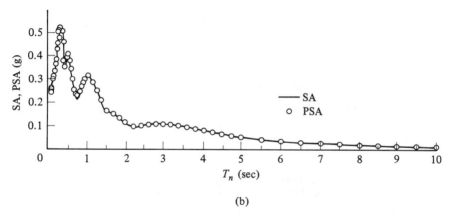

(b)

Figure 4.2   Response spectra and pseudo-response spectra ($\zeta = 0.05$) for N11°W component of Eureka, California, earthquake of December 21, 1954. (a) SV and PSV. (b) SA and PSA.

typical for strong-earthquake ground motion. In seeking to apply spectrum concepts to engineering design, we are not really concerned with the response of an ideal damped linear oscillator to the recorded motion of some past earthquake, but rather with the response of a real structure, existing or yet to be built, to an unknown earthquake that may occur in the future. For engineering purposes, PSA and SA may be used interchangeably.

Because of the exact mathematical relations between SD, PSV, and PSA, we can plot them on a three-way logarithmic plot much the same as we did for the dynamic response factors R_d, R_v, and R_a in Fig. 2.18. One family of curves will then give all three response spectra. Figure 4.3 shows the family of curves for the earthquake component of Fig. 4.1. This single graph shows the characteristics of the earthquake motion that are significant

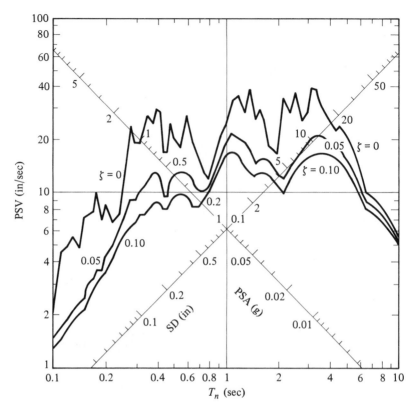

Figure 4.3 Three-way logarithmic response spectra for N11°W component of Eureka, California, earthquake of December 21, 1954.

in earthquake engineering. Knowing the natural period and the fraction of critical damping of a structure, we enter the spectrum plot at that value of period on the horizontal period scale, go up vertically to the curve for the correct fraction of critical damping, and then go over horizontally to the PSV scale to read the approximate maximum relative velocity, or go diagonally to the SD scale to read the "true" maximum relative displacement, or diagonally the other way to the PSA scale to read the maximum total acceleration, very close to true.

The response spectrum has proved so useful in strong-motion seismology and earthquake engineering that spectra for virtually all strong earthquakes that are recorded anywhere in the world are now computed and published soon after the events occur. Enough of them have been obtained to give us a reasonable idea of the kind of motion that is likely to be encountered in future strong earthquakes, and we are beginning to understand the influence of regional geology and local soil conditions on the response spectra.

4.4 RELATION TO SEISMIC BUILDING CODES

Seismic building codes can be portrayed in the same manner as response spectra. The Uniform Building Code employs a seismic coefficient C, which gives the lateral design force as a fraction of the weight of the building:

$$C = 1/15T_n^{1/2} \qquad \text{but } C \leq 0.12$$

The acceleration response spectrum SA, expressed in gravitational acceleration units, is the maximum force on an oscillator as a fraction of the weight of the oscillator. Hence, C and SA have comparable, although not identical, meanings. A direct comparison is clouded by such issues as factors of safety on working stresses and the differences between static and dynamic behavior. Figure 4.4 shows the UBC seismic coefficient C superimposed on the response spectra of Fig. 4.3. Note that the spectral accelerations are considerably greater than the code seismic coefficient over most of the period

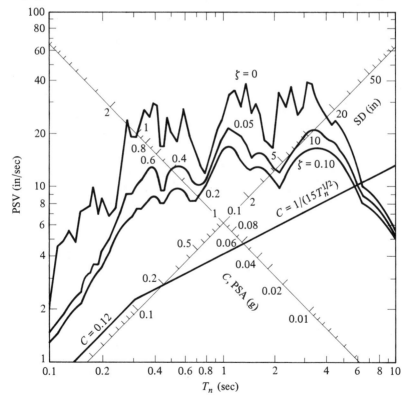

Figure 4.4   Building Code seismic coefficient and response spectra for N11°W component of Eureka, California, earthquake of December 21, 1954.

range. This earthquake caused damage but not great destruction. In terms of energy release, it was less than 1 percent as large as the great San Francisco earthquake of 1906.

Figure 4.5 shows the response spectra for the N–S component of the El Centro, California, earthquake of May 18, 1940, again with the building code seismic coefficient superimposed. The El Centro earthquake was significantly greater and more destructive than the Eureka earthquake, and the response spectra also are greater. Indeed, for three decades, the El Centro earthquake stood as the most intense ground motion ever recorded by a strong-motion accelerograph, although more severe ground motion had occurred without being so recorded.

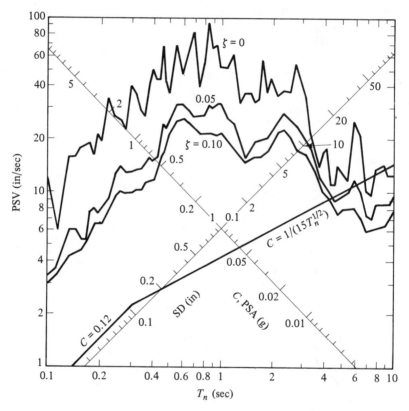

Figure 4.5 Building Code seismic coefficient and response spectra for N–S component of El Centro, California, earthquake of May 18, 1940.

PROBLEMS

4.1 Suppose the vertical cantilever of Problem 2.2 were subjected to the horizontal ground motion of that component of the Eureka earthquake characterized by the response spectra of Fig. 4.3. If damping were 5 percent of critical, what would be the maximum displacement of the mass and the maximum bending stress induced in the cantilever?

4.2 The ash hopper shown consists of a bin mounted on a rigid platform supported by four columns 20 ft long. The weight of the platform, bin, and contents is 150 kips, and may be taken as a point mass located 5 ft above the bottom of the platform. The columns are braced in the longitudinal direction, that is, normal to the plane of the paper, but are unbraced in the transverse direction.

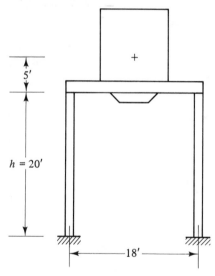

The column properties are

$$A = 20 \text{ in}^2$$
$$E = 29{,}500 \text{ ksi}$$
$$I = 1000 \text{ in}^4$$
$$S = 120 \text{ in}^3$$

Taking damping to be 5 percent of critical, find the maximum lateral displacement and the maximum stress in the columns due to gravity and the lateral motion of the El Centro earthquake component characterized by the response spectra of Fig. 4.5, acting in the transverse direction. Take the columns to be fixed at the base and rigidly connected to the rigid platform. Neglect axial deformation of the columns and P–Δ effects.

CHAPTER FIVE

Multidegree-of-Freedom Systems

5.1 INTRODUCTION

To this point, we have considered systems having a single degree of freedom, either damped or undamped. We have considered analytical methods of solving the equations of motion for linear systems, including the development of Duhamel's integral, which gave us a general solution for an arbitrary driving force. This proved useful in developing response-spectrum relations, but Duhamel's integral is difficult to evaluate and it is often more practical to attack the original differential equation of motion some other way. The response spectrum offered a convenient way of describing the extreme responses of all possible linear SDF systems to a given component of earthquake motion, thus enabling us to find the maximum response of a system merely by determining its frequency and damping and then picking the response from a chart.

Analytical procedures and response-spectrum techniques generally require that the system be linear, that is, that the differential equation of motion contain only the first powers of the displacement, velocity, and acceleration. Either material or geometric properties of the system could lead to a nonlinear differential equation of motion for which those procedures would fail. We were, however, able to develop numerical methods that would yield an approximate solution for either linear or nonlinear systems, and by choosing a suitably short time step and using adequate precision in the calculations, we could keep the error in the approximate solution almost arbitrarily small.

Numerical methods are applicable to multidegree-of-freedom (MDF) systems as well, either linear or nonlinear MDF systems, and we will explore them ahead of analytical methods. But first we must develop the differential equations of motion.

5.2 THE EQUATIONS OF MOTION

Let Fig. 5.1 represent a typical transverse frame of a building. For simplicity, we neglect the contributions of the partitions, fireproofing, and nonstructural elements to the lateral stiffness of the building, and we treat the frame as though the floors and roof were infinitely rigid compared with the columns. This is the often-used "shear-building" approximation. We also neglect axial deformation of the members, and we neglect the effect of axial force on the stiffness of the columns. The lateral stiffness of the building is then due entirely to the stiffness of the columns in bending. Figure 5.2 shows the forces on a distorted column.

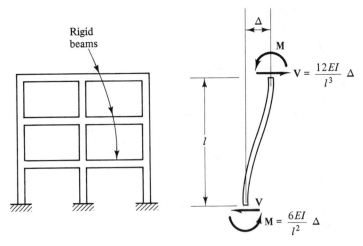

Figure 5.1 Idealized shearbuilding.   **Figure 5.2** Forces on distorted column.

The mass is distributed throughout the building, but we will treat it as though it were concentrated at the floor levels — the "lumped-mass" approximation. Each lumped mass comprises the entire mass of the building from midheight of the story below to midheight of the story above, structural and nonstructural components alike. We treat the building as though it were a weightless frame with lumped masses attached to it at the floor levels, as shown in Fig. 5.3. The masses are constrained to move only horizontally, and each is linked to the frame so that the horizontal inertia force enters into the equations of dynamic equilibrium, but there is no vertical motion of the mass.

Damping is elusive, here as in nearly all dynamic systems. We found for single-degree-of-freedom systems that velocity-proportional damping, that is, viscous damping, was mathematically convenient and led to vibration solutions that were consistent with the observed small-amplitude vibrations of some real systems. We portrayed the damping force as that force exerted by a dashpot attached to the mass, producing a force proportional to the velocity of the mass and opposed to the direction of the motion. The multidegree system complicates the concept, for the velocity-proportional damping force exerted on a mass could be taken as proportional to the velocity of that mass either relative to a fixed frame of reference, "absolute" damping, or relative to other masses in the system, "relative" damping. Without arguing the merits of one choice or the other, we will arbitrarily opt for relative damping, which we will portray as the effects of interfloor dashpots, each generating a damping force proportional to the velocity of the horizontal interfloor movement.

With these assumptions, we can write the equations of motion by considering free-body diagrams of the masses and writing the equations of dynamic equilibrium for each. From Fig. 5.3, let the three masses, numbered from bottom to top, be m_{11}, m_{22}, and m_{33}; let the three damping coefficients be d_1, d_2, d_3; and the three story stiffnesses, each representing the stiffness of all columns in the story, be s_1, s_2, and s_3. Further, let the dynamic lateral forces acting on the masses be $f_1(t)$, $f_2(t)$, and $f_3(t)$. Then Figure 5.4 shows free-body diagrams of the three masses in dynamic equilibrium.

Writing the equation of motion for each mass, we get a system of three linear differential equations of motion:

$$f_3(t) - m_{33}\ddot{u}_3 - d_3(\dot{u}_3 - \dot{u}_2) - s_3(u_3 - u_2) = 0$$

$$f_2(t) - m_{22}\ddot{u}_2 + d_3(\dot{u}_3 - \dot{u}_2) - d_2(\dot{u}_2 - \dot{u}_1) + s_3(u_3 - u_2)$$

$$- s_2(u_2 - u_1) = 0 \qquad (5.2.1)$$

$$f_1(t) - m_{11}\ddot{u}_1 + d_2(\dot{u}_2 - \dot{u}_1) - d_1\dot{u}_1 + s_2(u_2 - u_1) - s_1u_1 = 0$$

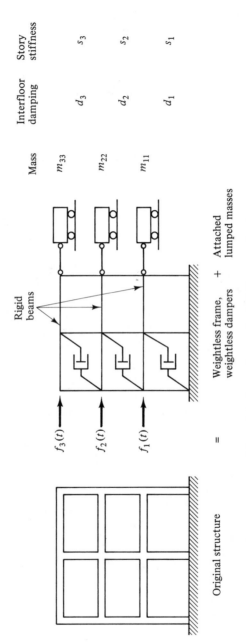

Figure 5.3 Lumped-mass shearbuilding approximation.

Mass	Interfloor damping	Story stiffness
m_{33}	d_3	s_3
m_{22}	d_2	s_2
m_{11}	d_1	s_1

Rigid beams

$f_3(t)$

$f_2(t)$

$f_1(t)$

=

Original structure

Weightless frame, weightless dampers   +   Attached lumped masses

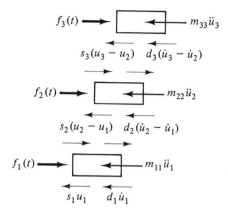

Figure 5.4 Dynamic forces on lumped masses.

We can rewrite these as

$$m_{11}\ddot{u}_1 + (d_1 + d_2)\dot{u}_1 - d_2\dot{u}_2 + (s_1 + s_2)u_1 - s_2u_2 = f_1(t)$$

$$m_{22}\ddot{u}_2 - d_2\dot{u}_1 + (d_2 + d_3)\dot{u}_2 - d_3\dot{u}_3 - s_2u_1$$

$$+ (s_2 + s_3)u_2 - s_3u_3 = f_2(t) \qquad (5.2.2)$$

$$m_{33}\ddot{u}_3 - d_3\dot{u}_2 + d_3\dot{u}_3 - s_3u_2 + s_3u_3 = f_3(t)$$

These are of the form

$$[M]\{\ddot{u}\} + [C]\{\dot{u}\} + [K]\{u\} = \{f(t)\} \qquad (5.2.3)$$

where

$$[M] = \begin{bmatrix} m_{11} & 0 & 0 \\ 0 & m_{22} & 0 \\ 0 & 0 & m_{33} \end{bmatrix} = \text{the inertia matrix} \qquad (5.2.4)$$

$$[C] = \begin{bmatrix} d_1 + d_2 & -d_2 & 0 \\ -d_2 & d_2 + d_3 & -d_3 \\ 0 & -d_3 & d_3 \end{bmatrix} = \text{the damping matrix} \qquad (5.2.5)$$

$$[K] = \begin{bmatrix} s_1 + s_2 & -s_2 & 0 \\ -s_2 & s_2 + s_3 & -s_3 \\ 0 & -s_3 & s_3 \end{bmatrix} = \text{the stiffness matrix} \qquad (5.2.6)$$

$$\{u\} = \begin{Bmatrix} u_1 \\ u_2 \\ u_3 \end{Bmatrix} = \text{the displacement vector} \qquad (5.2.7)$$

$$\{f\} = \begin{Bmatrix} f_1(t) \\ f_2(t) \\ f_3(t) \end{Bmatrix} = \text{the driving force vector} \qquad (5.2.8)$$

We identify the elements of a matrix by two subscripts, the first indicating the row position, numbered from top to bottom, and the second indicating the column position, numbered from left to right. Similarly, we identify the elements of a vector by a single subscript denoting the row position, again numbered from top to bottom.

The element m_{ij} in the inertia matrix $[M]$ is the inertia force at i corresponding to an acceleration $\ddot{u}_j = 1$, with all other $\ddot{u} = 0$.

The element c_{ij} in the damping matrix $[C]$ is the damping force at i corresponding to a velocity $\dot{u}_j = 1$, with all other $\dot{u} = 0$.

The element k_{ij} in the stiffness matrix $[K]$ is the restoring force at i corresponding to a displacement $u_j = 1$, with all other $u = 0$.

5.3 NUMERICAL EXAMPLE

Let the properties of the system be those shown in Fig. 5.5. With these properties, the structure has the following inertia, damping, and stiffness matrices, respectively,

$$[M] = \begin{bmatrix} 100 & 0 & 0 \\ 0 & 100 & 0 \\ 0 & 0 & 50 \end{bmatrix} \text{kips/g} \tag{5.3.1}$$

$$[C] = \begin{bmatrix} 1.0 & -0.5 & 0 \\ -0.5 & 1.0 & -0.5 \\ 0 & -0.5 & 0.5 \end{bmatrix} \text{kips/(in/sec)} \tag{5.3.2}$$

$$[K] = \begin{bmatrix} 250 & -100 & 0 \\ -100 & 150 & -50 \\ 0 & -50 & 50 \end{bmatrix} \text{kips/in} \tag{5.3.3}$$

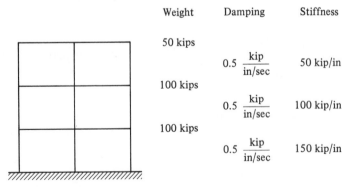

	Weight	Damping	Stiffness
	50 kips	$0.5\ \dfrac{\text{kip}}{\text{in/sec}}$	50 kip/in
	100 kips	$0.5\ \dfrac{\text{kip}}{\text{in/sec}}$	100 kip/in
	100 kips	$0.5\ \dfrac{\text{kip}}{\text{in/sec}}$	150 kip/in

Figure 5.5 System properties for numerical example.

The matrices of Eqs. (5.3.1) through (5.3.3) may be obtained by inserting the properties shown in Fig. 5.5 into Eqs. (5.2.4) through (5.2.6). Alternatively, the element definitions may be used. The stiffness matrix, for example, may be obtained by first imposing a unit displacement $u_1 = 1$. The forces required to hold the system in static equilibrium in this configuration are the stiffness elements k_{11}, k_{21}, and k_{31}, the first column of matrix $[K]$. Then imposing a unit displacement $u_2 = 1$ gives the second column of $[K]$, and imposing $u_3 = 1$ gives the third column of $[K]$, as indicated in Fig. 5.6.

$$[K] = \begin{bmatrix} k_{11} & k_{12} & k_{13} \\ k_{21} & k_{22} & k_{23} \\ k_{31} & k_{32} & k_{33} \end{bmatrix}$$

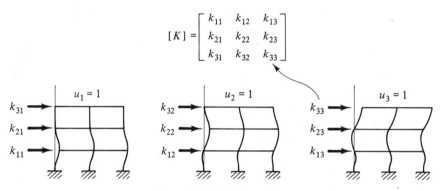

Figure 5.6   Stiffness matrix.

5.4 PROPERTIES OF SYSTEM MATRICES

Several properties of the system matrices may be observed in $[M]$, $[C]$, and $[K]$, in the previous section, and are useful to bear in mind to avoid potential mistakes.

First, $[M]$, $[C]$, and $[K]$ are symmetrical matrices. In this case, $[M]$ is a diagonal matrix, but this is not always true. The symmetry of $[K]$ follows from Maxwell's reciprocal theorem, which is based on the principle of superposition and strain-energy considerations. Comparable arguments using energy dissipated in viscous damping instead of strain energy can be used to establish the symmetry of $[C]$, and kinetic-energy considerations lead to the symmetry of $[M]$.

Second, the diagonal elements of $[M]$ and $[K]$ are always positive, due to the positive definite character of strain energy and kinetic energy. Similarly, the diagonal elements of $[C]$ are always nonnegative because any velocity must be accompanied by either dissipation of energy in viscous damping, or, at best, by no change in energy — damping cannot increase the energy of the system. The possibility of part of the system being undamped accounts for the possibility of zero diagonal elements.

Third, the largest element of [M] will always be on the diagonal, as will the largest element of [K]. Some off-diagonal elements may be larger than some diagonal elements, but the largest element in the entire matrix will always be on the diagonal.

5.5 NUMERICAL SOLUTION

Numerical methods may be employed to solve the differential equations of motion for a multidegree dynamic system. The procedure is straightforward, but because the equations are ordinarily coupled, as in this case, they must be solved simultaneously. Any of the numerical methods of Chapter 3 may be used except the modified linear-acceleration method, which was specifically tailored for the SDF linear system. They are applicable to either linear or nonlinear MDF systems.

Take the three-story frame example of Fig. 5.5, and let the driving force be the single sine pulse of Fig. 5.7 acting at the second-floor level. The equations of motion are

$$\{\ddot{u}\} = [M]^{-1}\{\{f\} - [C]\{\dot{u}\} - [K]\{u\}\} \tag{5.5.1}$$

Because the inertia matrix [M] is diagonal, we can write Eq. (5.5.1) as

$$\ddot{u}_i = (f_i - \sum_j c_{ij}\dot{u}_j - \sum_j k_{ij}u_j)/m_{ii} \qquad i = 1, 2, \ldots, n \tag{5.5.2}$$

If [M] were not diagonal, we would need to use the less convenient form:

$$\ddot{u}_i = \sum_j m_{ij}^{-1}\left(f_j - \sum_p c_{jp}\dot{u}_p - \sum_p k_{jp}u_p\right) \qquad i = 1, 2, \ldots, n \tag{5.5.3}$$

$f_2 \longrightarrow$

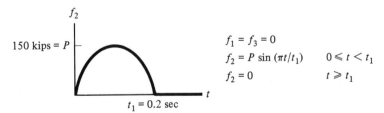

f_2

150 kips = P

$t_1 = 0.2$ sec

$f_1 = f_3 = 0$
$f_2 = P \sin(\pi t/t_1) \qquad 0 \le t < t_1$
$f_2 = 0 \qquad\qquad\qquad t \ge t_1$

Figure 5.7 Driving force for numerical example.

in which m_{ij}^{-1} denotes the ij element of $[M]^{-1}$.

Program 5.1 solves the equations for the case of diagonal $[M]$ using the Runge–Kutta method.

Program 5.1: Runge–Kutta Program for the MDF System

```
10     GOSUB 1000           ' Get data
20     GOSUB 2000           ' Start solution
30     WHILE T<TEND
40       GOSUB 5000         ' Advance one time step
50     WEND
60     CLOSE #1
70     END

1000                       ' Housekeeping instructions and data
1010 DEFINT I,J,N
1020 F3$ = "####.###"
1030 F4$ = " ####.####"
1040 GRAV=386.088
1050 PI=3.141593
1060 CLS: INPUT "Output device or filespec:  ", O$
1070 OPEN O$ FOR OUTPUT AS #1
1080 READ N
1090 DIM W(N),C(N,N),K(N,N),U(N),V(N),A(N)
1100 DIM OLDU(N),OLDV(N),K1(N),K2(N),K3(N),K4(N)
1110 FOR I=1 TO N
1120   READ W(I)
1130 NEXT I
1140 FOR I=1 TO N
1150   FOR J=1 TO N
1160     READ C(I,J)
1170   NEXT J
1180 NEXT I
1190 FOR I=1 TO N
1200   FOR J=1 TO N
1210     READ K(I,J)
1220   NEXT J
1230 NEXT I
1240 READ P,T1,H,TEND
1250 PRINT#1, "      T   ";
1260 FOR I=1 TO N
1270   PRINT#1, "     U(";I;")";
1280 NEXT I
1290 PRINT#1,
1300 RETURN
```

```
1500                          ' Data for example problem
1510 DATA    3
1520 DATA 100,  100,    50
1530 DATA    1, -0.5,    0, -0.5,    1, -0.5, 0, -0.5, 0.5
1540 DATA 250, -100,    0, -100, 150,  -50, 0,  -50,  50
1550 DATA 150, 0.2, 0.01,  0.7

2000                          ' Starter instructions
2010 T = 0
2020 FOR I=1 TO N
2030   U(I)=0
2040   V(I)=0
2050 NEXT I
2060 GOSUB 3000
2070 GOSUB 4000
2080 RETURN

3000                          ' Acceleration subroutine
3010 FOR I=1 TO N
3020   IF I=2 AND T<T1 THEN F=P*SIN(PI*T/T1) ELSE F=0
3030   D=0
3040   Q=0
3050   FOR J=1 TO N
3060     D=D+C(I,J)*V(J)
3070     Q=Q+K(I,J)*U(J)
3080   NEXT J
3090 A(I)=(F-D-Q)*GRAV/W(I)
3100 NEXT I
3110 RETURN

4000                          ' Output subroutine
4010 PRINT#1, USING F3$; T;
4020 FOR I=1 TO N
4030   PRINT#1, USING F4$; U(I);
4040 NEXT I
4050 PRINT#1,
4060 RETURN

5000                          ' Runge-Kutta fourth-order subroutine
5010 T=T+H/2
5020 FOR I=1 TO N
5030   OLDU(I)=U(I)
5040   OLDV(I)=V(I)
5050   K1(I)=H*A(I)
5060   U(I)=OLDU(I)+H*(OLDV(I)/2+K1(I)/8)
5070   V(I)=OLDV(I)+K1(I)/2
5080 NEXT I
```

```
5090    GOSUB 3000          ' Acceleration
5100 FOR I=1 TO N
5110    K2(I)=H*A(I)
5120    V(I)=OLDV(I)+K2(I)/2
5130 NEXT I
5140 GOSUB 3000             ' Acceleration
5150 T=T+H/2
5160 FOR I=1 TO N
5170    K3(I)=H*A(I)
5180    U(I)=OLDU(I)+H*(OLDV(I)+K3(I)/2)
5190    V(I)=OLDV(I)+K3(I)
5200 NEXT I
5210 GOSUB 3000             ' Acceleration
5220 FOR I=1 TO N
5230    K4(I)=H*A(I)
5240    U(I)=OLDU(I)+H*(OLDV(I)+(K1(I)+K2(I)+K3(I))/6)
5250    V(I)=OLDV(I)+(K1(I)+2*(K2(I)+K3(I))+K4(I))/6
5260 NEXT I
5270    GOSUB 3000          ' Acceleration
5280    GOSUB 4000          ' Print
5290    RETURN
```

The program is a direct adaptation of Program 3.5, except that in the Runge–Kutta subroutine, statements 5000–5290, each partial time step is completed for all N equations before the next partial time step is undertaken. Thus, all N of the k_1's are calculated and all N of the u_i's and $\dot{u}_i$'s are advanced, then the acceleration subroutine is called to calculate the N $\ddot{u}_i$'s, then all N k_2's are calculated, all N $\dot{u}_i$'s are advanced, etc.

Program 5.2 solves the same example using the linear-acceleration method.

Program 5.2: Linear-Acceleration Program for the MDF System

A program to solve the same problem by the linear-acceleration method may be derived by modifying Program 5.1 as follows:

1. Replace statement 1100, for the variables dimensioned therein are not used, but three different subscripted variables are needed:

   ```
   1100 DIM UX(N), VX(N), AX(N)
   ```

2. Insert statements to read a convergence test epsilon:

   ```
   1245 READ EPSILON
   1555 DATA 0.001
   ```

3. Replace the Runge–Kutta subroutine, statements 5000–5290, with the following linear-acceleration subroutine:

```
5000                                    ' Linear-acceleration subroutine
5010 FOR I=1 TO N
5020    UX(I) = U(I)+H*(V(I)+H*A(I)/3)
5030    VX(I) = V(I)+H*A(I)/2
5040 NEXT I
5050 T=T+H
5060 CHANGE=1+EPSILON
5070 WHILE CHANGE>EPSILON
5080    FOR I=1 TO N
5090       AX(I)=A(I)
5100       U(I)=UX(I)+H*H*A(I)/6
5110       V(I)=VX(I)+H*A(I)/2
5120    NEXT I
5130    GOSUB 3000              ' Acceleration
5140    CHANGE=0
5150    FOR I=1 TO N
5160       DA=ABS(AX(I)-A(I))
5170       IF DA>CHANGE THEN CHANGE=DA
5180    NEXT I
5190 WEND
5200 GOSUB 4000                 ' Print
5210 RETURN
```

Figure 5.8 shows how the three displacements vary with time. The second floor, which is where the driving force is applied, starts to move

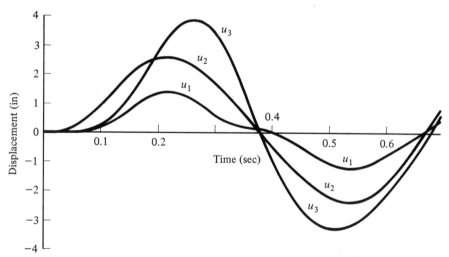

Figure 5.8 Dynamic response for numerical example.

first. The first and third floors are soon drawn along, and the third floor overtakes the second. After the force pulse ends at $t = 0.20$ sec, the system is in damped free vibration and oscillates more or less sinusoidally, with some higher-frequency components apparent.

5.6 ANALYTICAL SOLUTION

An analytical solution of the MDF equation, Eq. (5.2.3), as it stands is not feasible; however, we can invoke the concept of normal modes of vibration to recast the equation in an alternative form that is amenable to analysis.

Our experience with common vibrating systems suggests the existence of modes of vibration. Pluck a taut wire and it vibrates mainly in the shape of a single wave. Careful observation reveals other waves of shorter wavelength superimposed on the primary wave. Moreover, the dominant tone remains the same unless we change the tension or the length of the wire. Plucking it harder changes the volume, but not the primary tone. And, if we listen carefully, we can hear overtones in addition to the primary tone. These single and multiple waves are the fundamental and higher modes of vibration, respectively — the fundamental tone and overtones. Similarly, if we take a multiple pendulum, say a string with two or more attached weights, we will find that we can set it swinging in a shape that remains the same, that is, with the displacements of the weights always proportional to one another, with only the amplitude changing. With some experimentation, we can find a second shape, or possibly even a third if there are three or more weights, so that if the pendulum is displaced in that shape and released, it will oscillate in that shape, with only the amplitude changing.

This suggests an analytical approach to the MDF vibration problem analogous to the separation of variables in classical mathematical analysis. Let us assume that in free vibration the undamped MDF system can oscillate in a fixed shape, with only the amplitude varying with time. Specifically, let us assume that in free vibration, the displacement vector $\{u(t)\}$ can be represented as

$$\{u(t)\} = \{\phi\}z(t) \qquad (5.6.1)$$

in which $\{\phi\}$ is a dimensionless constant vector with at least one nonzero element, and z is the generalized displacement, a function of time. The nonzero restriction on $\{\phi\}$ serves merely to preclude the trivial case of no vibration at all. The vector $\{\phi\}$ may contain zero elements, but at least one element must be nonzero. Thus far *this is only a hypothesis*. We now proceed to explore its consequences.

The equations of motion for the undamped MDF linear system in free vibration are

$$[M]\{\ddot{u}\} + [K]\{u\} = 0 \tag{5.6.2}$$

From Eq. (5.6.1), we get

$$\{\ddot{u}\} = \{\phi\}\ddot{z}(t) \tag{5.6.3}$$

Now we put Eqs. (5.6.1) and (5.6.3) into Eq. (5.6.2) to get

$$[M]\{\phi\}\ddot{z} + [K]\{\phi\}z = 0 \tag{5.6.4}$$

For any given system, $[M]$ and $[K]$ are constant matrices, and by hypothesis, $\{\phi\}$ is a constant vector. Thus, Eq. (5.6.4) takes the form

$$\ddot{z} + (\text{constant})\, z = 0 \tag{5.6.5}$$

We encountered this earlier in Eq. (2.1.10) when we considered the single-degree-of-freedom undamped system. The constant then was ω_n^2, the square of the undamped natural frequency. The equation took the form

$$\ddot{z} + \omega_n^2 z = 0 \tag{5.6.6}$$

for which one solution is

$$z = A \sin \omega_n t \tag{5.6.7}$$

Thus, if our hypothesis, Eq. (5.6.1), is valid, then the displacement of the system in free vibration may take the form

$$\{u\} = \{\phi\}A \sin \omega t \tag{5.6.8}$$

from which

$$\{\ddot{u}\} = -\omega^2\{\phi\}A \sin \omega t \tag{5.6.9}$$

Put Eqs. (5.6.8) and (5.6.9) into Eq. (5.6.2) to get

$$(-\omega^2[M]\{\phi\} + [K]\{\phi\})A \sin \omega t = 0 \tag{5.6.10}$$

For a nontrivial solution to exist, A and ω must both be nonzero; hence,

$$[[K] - \omega^2[M]]\{\phi\} = 0 \tag{5.6.11}$$

By hypothesis, at least one element of $\{\phi\}$ is nonzero. Therefore, Eq. (5.6.11) requires that the matrix $[[K] - \omega^2[M]]$ be singular, which requires that it have a zero determinant. Thus, the hypothesis of Eq. (5.6.1) leads us to the frequency equation

$$|[K] - \omega^2[M]| = 0 \tag{5.6.12}$$

5.7 MODES AND FREQUENCIES

For the numerical example of Sec. 5.3, Eq. (5.6.12) is

$$\|[K] - \omega^2[M]\| = \begin{vmatrix} 250 - 100\omega^2 & -100 & 0 \\ -100 & 150 - 100\omega^2 & -50 \\ 0 & -50 & 50 - 50\omega^2 \end{vmatrix} = 0$$

$$(5.7.1)$$

In Eq. (5.7.1), the stiffness matrix $[K]$ is in units of kips/in and the inertia matrix $[M]$ is in units of kips/g. Therefore, ω^2 will be in units of $in^{-1}g$, for $[K]$ and $\omega^2[M]$ must be dimensionally alike.

Expanding the determinant of Eq. (5.7.1) gives us a polynomial

$$-500,000\omega^6 + 2,500,000\omega^4 - 3,125,000\omega^2 + 750,000 = 0 \qquad (5.7.2)$$

or, after dividing through by the lead coefficient,

$$\omega^6 - 5\omega^4 + 6.25\omega^2 - 1.5 = 0 \qquad (5.7.3)$$

Descartes' rule of signs tells us that Eq. (5.7.3) has either three or one positive root and no negative root. From the theory of equations, we know that the sum of the three ω^2's is 5, the negative of the coefficient of the next highest power. In general, the frequency equation will be a polynomial of degree $2N$ with the odd powers missing, perhaps better described as a polynomial of degree N in ω^2. The positive definite character of $[K]$ and $[M]$ for any stable linear dynamic system ensures that all N roots will be real and positive, although not necessarily distinct.

Newton's method is one technique for finding the roots of a polynomial, or, for that matter, the roots of any continuous and differentiable function. To find a root of the function $f(x) = 0$ by Newton's method, we start with an approximate root x_0, compute an improved approximation $x_1 = x_0 - f(x_0)/f'(x_0)$, use this to get a better x_2, use x_2 to get x_3, and so on. In general, the iteration is

$$x_{n+1} = x_n - f(x_n)/f'(x_n) \qquad (5.7.4)$$

Figure 5.9 illustrates the procedure, and Table 5.1 shows the iteration for the highest root.

We number the frequencies in increasing order, $\omega_1 \leq \omega_2 \leq \omega_3$. For this case, they are

$$\omega_1^2 = 0.31386 \ in^{-1}g = \ \ 121.18 \ sec^{-2}; \quad or \ \omega_1 = 11.008 \ rad/sec$$
$$\omega_2^2 = 1.5 \ in^{-1}g \ \ \ \ \ = \ \ 579.13 \ sec^{-2}; \quad or \ \omega_2 = 24.065 \ rad/sec$$
$$\omega_3^2 = 3.18614 \ in^{-1}g = 1230.13 \ sec^{-2}; \quad or \ \omega_3 = 35.073 \ rad/sec$$

$$(5.7.5)$$

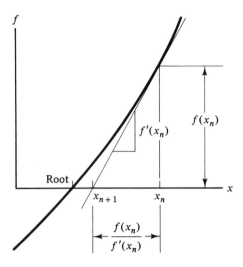

Figure 5.9 Newton's method.

TABLE 5.1 Newton's Method Iteration

Approximate ω^2	Polynomial $f(\omega^2)$	Derivative $f'(\omega^2)$	Improved approx. $\omega^2 - f(\omega^2)/f'(\omega^2)$
5	29.75	31.25	4.048
4.048	8.20024	14.92891	3.49871
3.49871	1.98972	7.98586	3.24596
3.24596	0.32573	5.43331	3.18961
3.18961	0.01685	4.87472	3.18615
3.18615	0.00005	4.84317	3.18614
3.18614	0.00000		

To this point, we can conclude that for our hypothetical free vibration of the system, Eq. (5.6.1), to be valid, Eq. (5.6.6) must also be true, and ω^2 must be a root of the frequency equation, Eq. (5.6.12). Thus, for our numerical example, the free vibration must be a sinusoidal oscillation at one of three distinct frequencies.

We return now to Eq. (5.6.11) to determine $\{\phi\}$. We cannot determine it absolutely, for any multiple of a solution $\{\phi\}$ would also be a solution. We can, however, assign an arbitrary value to one element of $\{\phi\}$ and solve for the other elements in terms of that one. For the first frequency, for example, we put the value of ω_1^2 from Eq. (5.7.5) into Eq. (5.6.11) to get

$$[[K] - \omega_1^2[M]]\{\phi\} = \begin{bmatrix} 218.614 & -100 & 0 \\ -100 & 118.614 & -50 \\ 0 & -50 & 34.307 \end{bmatrix} \begin{Bmatrix} \phi_1 \\ \phi_2 \\ \phi_3 \end{Bmatrix} = \begin{Bmatrix} 0 \\ 0 \\ 0 \end{Bmatrix}$$

$$(5.7.6)$$

Arbitrarily, set $\phi_3 = 1$. Then we may solve any two of the three row equations in Eq. (5.7.6) for ϕ_2 and ϕ_1 and use the remaining row equation as a check. The solution gives us a vector $\{\phi\}_1$, the subscript indicating that it is associated with the frequency ω_1. In the same manner, we can find a vector $\{\phi\}_2$ associated with the frequency ω_2, and a $\{\phi\}_3$ associated with ω_3. Thus, we obtain the three mode shapes and frequencies indicated in Table 5.2. These are the three modes in which free vibration of our undamped MDF system may occur.

In Table 5.2, we normalized the $\{\phi\}$'s by making the largest element 1. There are other valid ways to normalize modes, for example, so that $\{\phi\}^T\{\phi\} = 1$ or so that $\{\phi\}^T[M]\{\phi\} = 1$. Some are mathematically more elegant than ours, but none is simpler.

What we have done is equivalent to solving the classical eigenvalue problem in mathematics. If we had premultiplied the terms of Eq. (5.6.11) by $[M]^{-1}$, it would have become

$$[[M]^{-1}[K]]\{\phi\} = \omega^2\{\phi\} \tag{5.7.7}$$

which is the eigenvalue problem. The ω^2's are the eigenvalues of $[[M]^{-1}[K]]$ and the $\{\phi\}$'s are its eigenvectors. This does not quite fit the standard eigenvalue subroutines usually found in computing center libraries, however, for those ordinarily require that the matrix be symmetrical, and while both $[M]$ and $[K]$ are symmetrical, $[[M]^{-1}[K]]$ is not. The symmetry requirement is imposed for computer subroutines because a symmetrical matrix has real ei-

TABLE 5.2 Mode Shapes and Frequencies

Mode number, p	1	2	3
ω_p^2 (sec^{-2})	121.18	579.13	1230.1
Frequency ω_p (rad/sec)	11.008	24.065	35.073
Period T_p (sec)	0.5708	0.2611	0.1791
Mode shape $\{\phi\}_p$	$\begin{Bmatrix} 0.31386 \\ 0.68614 \\ 1 \end{Bmatrix}$	$\begin{Bmatrix} -0.5 \\ -0.5 \\ 1 \end{Bmatrix}$	$\begin{Bmatrix} 1 \\ -0.68614 \\ 0.31386 \end{Bmatrix}$

genvalues, whereas the eigenvalues of an unsymmetrical matrix may be complex. The eigenvalues of the vibration-frequency problem are always real and the problem can be cast in the form of the standard symmetrical-matrix eigenvalue problem, but we will not consider that here.

The procedure we have undertaken, finding the frequencies by expanding the frequency determinant into a polynomial and finding its roots, is impractical for any system with more than a few degrees of freedom. Indeed, one mathematical technique for finding the roots of a high-order polynomial is the inverse of what we have done — in effect, find a linear MDF system having frequencies equal to the roots of the polynomial, and then find the frequencies by other means. We will consider more practical means for finding frequencies and mode shapes, but first we need to complete the circle and establish the validity of our original hypothesis, and that requires additional properties of the modes.

5.8 MODAL ORTHOGONALITY

For mode 1, we have, from Eq. (5.6.11),

$$[K]\{\phi\}_1 = \omega_1^2[M]\{\phi\}_1 \tag{5.8.1}$$

Premultiply both sides of Eq. (5.8.1) by $\{\phi\}_2^T$ to get

$$\{\phi\}_2^T[K]\{\phi\}_1 = \omega_1^2\{\phi\}_2^T[M]\{\phi\}_1 \tag{5.8.2}$$

Similarly, for mode 2, Eq. (5.6.11) gives us

$$[K]\{\phi\}_2 = \omega_2^2[M]\{\phi\}_2 \tag{5.8.3}$$

Premultiply both sides of Eq. (5.8.3) by $\{\phi\}_1^T$ to get

$$\{\phi\}_1^T[K]\{\phi\}_2 = \omega_2^2\{\phi\}_1^T[M]\{\phi\}_2 \tag{5.8.4}$$

The expressions on both sides of Eq. (5.8.2) are scalar quantities. Thus,

$$\{\phi\}_2^T[K]\{\phi\}_1 = (\{\phi\}_2^T[K]\{\phi\}_1)^T = \{\phi\}_1^T[K]^T\{\phi\}_2 \tag{5.8.5}$$

But $[K]$ is symmetrical, so $[K] = [K]^T$ and, therefore,

$$\{\phi\}_2^T[K]\{\phi\}_1 = \{\phi\}_1^T[K]\{\phi\}_2 \tag{5.8.6}$$

Similar arguments lead to the conclusion that

$$\{\phi\}_2^T[M]\{\phi\}_1 = \{\phi\}_1^T[M]\{\phi\}_2 \tag{5.8.7}$$

and when we put Eqs. (5.8.6) and (5.8.7) into Eq. (5.8.2), we get

$$\{\phi\}_1^T[K]\{\phi\}_2 = \omega_1^2\{\phi\}_1^T[M]\{\phi\}_2 \tag{5.8.8}$$

Now subtract Eq. (5.8.8) from Eq. (5.8.4) to get

$$0 = (\omega_2^2 - \omega_1^2)\{\phi\}_1^T[M]\{\phi\}_2 \tag{5.8.9}$$

But $\omega_2^2 \neq \omega_1^2$, so

$$\{\phi\}_1^T[M]\{\phi\}_2 = 0 \tag{5.8.10}$$

and this in Eq. (5.8.8) gives

$$\{\phi\}_1^T[K]\{\phi\}_2 = 0 \tag{5.8.11}$$

Equations (5.8.10) and (5.8.11) are the orthogonality relations for modes 1 and 2. In general,

$$\left.\begin{array}{r}\{\phi\}_p^T[M]\{\phi\}_r = 0 \\[2ex] \text{and} \qquad\qquad\qquad\qquad\qquad \\[2ex] \{\phi\}_p^T[K]\{\phi\}_r = 0\end{array}\right\} \quad \text{for } p \neq r \tag{5.8.12}$$

We have only established Eq. (5.8.12) for $\omega_p \neq \omega_r$. If there is a repeated frequency, multiple $\{\phi\}$'s can be found that satisfy the orthogonality relations for that frequency, and Eq. (5.8.12) then remains valid for all modes.

5.9 A GEOMETRIC INTERPRETATION

We may illustrate the significance of modal orthogonality by considering the displacements u_1, u_2, and u_3 as taking place along the coordinate axes in a three-dimensional space. Any displaced configuration of the system $\{u\}$ then corresponds uniquely to a position vector in that space. The three mode shapes $\{\phi\}_1$, $\{\phi\}_2$, and $\{\phi\}_3$ are also unique position vectors in that space, as shown in Fig. 5.10.

In a three-dimensional Euclidean space, two vectors $\{a\}$ and $\{b\}$ are perpendicular (orthogonal) if their scalar product is zero, that is, if $\{a\}^T\{b\} = 0$. Here we have vectors $\{\phi\}_1$, $\{\phi\}_2$, and $\{\phi\}_3$ that are orthogonal with respect to weighting factors $[M]$ and $[K]$, that is, $\{\phi\}_p^T[M]\{\phi\}_r = 0$ and $\{\phi\}_p^T[K]\{\phi\}_r = 0$ for $p \neq r$. Any displaced position of the dynamic system corresponds uniquely to a position vector in the u_1–u_2–u_3 space. Just as we use unit vectors in the u_1, u_2, u_3 directions as the basis for a coordinate system, we may use the three "unit" vectors $\{\phi\}_1$, $\{\phi\}_2$, and $\{\phi\}_3$ as the basis for an alternate coordinate system. The orthogonality properties guarantee that

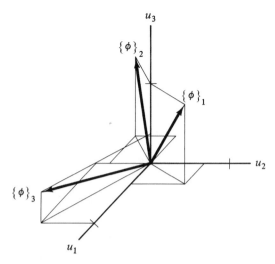

Figure 5.10 Mode-shape vectors in u space.

the N $\{\phi\}$'s for an N-degree-of-freedom system will span the N-dimensional space. The coordinate transformation is

$$\{u\} = \{\phi\}_1 q_1 + \{\phi\}_2 q_2 + \{\phi\}_3 q_3 \tag{5.9.1}$$

where q_1, q_2, and q_3 are the components of $\{u\}$ in the $\{\phi\}_1$, $\{\phi\}_2$, and $\{\phi\}_3$ directions, respectively. The q's are functions of time; the $\{\phi\}$'s are constants.

5.10 CHANGING COORDINATES

With n degrees of freedom, the coordinate transformation of Eq. (5.9.1) becomes

$$\{u\} = \sum_{p=1}^{n} \{\phi\}_p q_p \tag{5.10.1}$$

Sometimes we need to accomplish the inverse transformation, that is, convert from u coordinates to q coordinates. If we knew all of the modes, we could, at least in theory, assemble all n of the mode vectors $\{\phi\}_p$ into a matrix, invert it, and rewrite Eq. (5.10.1) the other way around. However, this would require knowing all of the modes and inverting a potentially large matrix, either of which could pose difficulties. We need a process that circumvents these requirements. Orthogonality provides the key.

Premultiply each term of Eq. (5.10.1) by $\{\phi\}_r^T[M]$ to get

$$\{\phi\}_r^T[M]\{u\} = \{\phi\}_r^T[M]\sum_{p=1}^{n}\{\phi\}_p q_p$$
$$= \{\phi\}_r^T[M]\{\phi\}_1 q_1 + \ldots + \{\phi\}_r^T[M]\{\phi\}_r q_r \qquad (5.10.2)$$
$$+ \ldots + \{\phi\}_r^T[M]\{\phi\}_n q_n$$

The orthogonality relations of Eq. (5.8.12) make every term on the right side of Eq. (5.10.2) zero except term r, leaving

$$\{\phi\}_r^T[M]\{u\} = \{\phi\}_r^T[M]\{\phi\}_r q_r$$

from which

$$q_r = \frac{\{\phi\}_r^T[M]\{u\}}{\{\phi\}_r^T[M]\{\phi\}_r} \qquad r = 1, 2, \ldots, n \qquad (5.10.3)$$

To illustrate, suppose the example structure of Fig. 5.5 were displaced 1 in at all floors, that is, $u_1 = u_2 = u_3 = 1$ in. What would be the corresponding displacements in q coordinates? With the mode shapes of Table 5.2, Eq. (5.10.3) gives us

$$q_1 = \{\phi\}_1^T[M]\{u\}/(\{\phi\}_1^T[M]\{\phi\}_1 = \quad 1.403 \text{ in}$$
$$q_2 = \{\phi\}_2^T[M]\{u\}/(\{\phi\}_2^T[M]\{\phi\}_2 = -0.500 \text{ in} \qquad (5.10.4)$$
$$q_3 = \{\phi\}_3^T[M]\{u\}/(\{\phi\}_3^T[M]\{\phi\}_3 = \quad 0.309 \text{ in}$$

Observe that each q_p is determined using only the properties of mode p. Figure 5.11 shows the results.

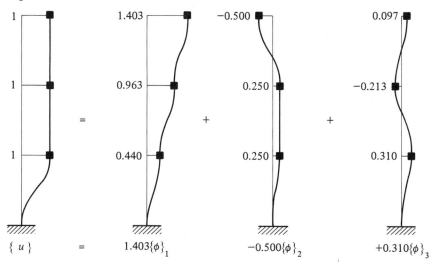

Figure 5.11   Coordinate transformation.

5.11 MODAL EQUATIONS OF MOTION

The equation of motion for forced vibration of an undamped system, from Eq. (5.2.3), is

$$[M]\{\ddot{u}\} + [K]\{u\} = \{f\} \tag{5.11.1}$$

Let us now put the coordinate transformation of Eq. (5.10.1) into Eq. (5.11.1) to get

$$
\begin{aligned}
[M]\{\phi\}_1\ddot{q}_1 + [M]\{\phi\}_2\ddot{q}_2 + [M]\{\phi\}_3\ddot{q}_3 \\
+ [K]\{\phi\}_1q_1 + [K]\{\phi\}_2q_2 + [K]\{\phi\}_3q_3 = \{f\}
\end{aligned}
\tag{5.11.2}
$$

Now we premultiply each term of Eq. (5.11.2) by $\{\phi\}_1^T$ to get

$$
\begin{aligned}
\{\phi\}_1^T[M]\{\phi\}_1\ddot{q}_1 + \{\phi\}_1^T[M]\{\phi\}_2\ddot{q}_2 + \{\phi\}_1^T[M]\{\phi\}_3\ddot{q}_3 \\
+ \{\phi\}_1^T[K]\{\phi\}_1q_1 + \{\phi\}_1^T[K]\{\phi\}_2q_2 + \{\phi\}_1^T[K]\{\phi\}_3q_3 = \{\phi\}_1^T\{f\}
\end{aligned}
\tag{5.11.3}
$$

By virtue of the orthogonality relations, Eq. (5.8.12), the coefficients of $\ddot{q}_2$, $\ddot{q}_3$, q_2, and q_3 in Eq. (5.11.3) are all zero, which leaves

$$\{\phi\}_1^T[M]\{\phi\}_1\ddot{q}_1 + \{\phi\}_1^T[K]\{\phi\}_1q_1 = \{\phi\}_1^T\{f\} \tag{5.11.4}$$

or

$$\ddot{q}_1 + \frac{\{\phi\}_1^T[K]\{\phi\}_1}{\{\phi\}_1^T[M]\{\phi\}_1}q_1 = \frac{\{\phi\}_1^T\{f\}}{\{\phi\}_1^T[M]\{\phi\}_1} \tag{5.11.5}$$

or

$$\ddot{q}_1 + \omega_1^2 q_1 = \frac{\{\phi\}_1^T\{f\}}{\{\phi\}_1^T[M]\{\phi\}_1} \tag{5.11.6}$$

Similarly, premultiplying the terms of Eq. (5.11.2) by $\{\phi\}_2^T$ and $\{\phi\}_3^T$ lead to

$$\ddot{q}_2 + \omega_2^2 q_2 = \frac{\{\phi\}_2^T\{f\}}{\{\phi\}_2^T[M]\{\phi\}_2} \tag{5.11.7}$$

and

$$\ddot{q}_3 + \omega_3^2 q_3 = \frac{\{\phi\}_3^T\{f\}}{\{\phi\}_3^T[M]\{\phi\}_3} \tag{5.11.8}$$

Equations (5.11.6) to (5.11.8) are the three modal equations of motion for forced vibration of the undamped system. They are completely uncoupled and can be solved separately. Each modal equation is identical in form to the equation of motion for forced vibration of an undamped linear SDF system, Eq. (1.6.2) with the damping term removed, so any of the analytical or numerical methods developed in earlier chapters are applicable. Note that the equation for q_1 involves only the properties of the first mode. Thus, if we know the first mode, we can write the equation for q_1 and solve it without even knowing the other modes. We convert the q's to u coordinates by means of Eq. (5.10.1).

Now we can validate the hypothesis of Eq. (5.6.1). The first-mode equation for free vibration is, from Eq. (5.11.6),

$$\ddot{q}_1 + \omega_1^2 q_1 = 0 \qquad (5.11.9)$$

for which one valid solution is

$$q_1 = A \sin \omega_1 t \qquad (5.11.10)$$

Equation (5.10.1) transforms this to u coordinates, giving us

$$\{u\} = \{\phi\}_1 q_1 = \{\phi\}_1 A \sin \omega_1 t \qquad (5.11.11)$$

which confirms our original hypothesis that such a free vibration could occur.

5.12 EFFECTIVE MASS AND STIFFNESS

We can write the modal differential equations of motion for our example system in a form comparable to Eq. (1.6.1) with the damping term deleted:

$$[M^*]\{\ddot{q}\} + [K^*]\{q\} = \{f^*(t)\} \qquad (5.12.1)$$

where the elements of the effective mass and stiffness matrices $[M^*]$ and $[K^*]$, respectively, are

$$m_{ij}^* = \{\phi\}_i^T [M] \{\phi\}_j \qquad (5.12.2)$$

and

$$k_{ij}^* = \{\phi\}_i^T [K] \{\phi\}_j \qquad (5.12.3)$$

and the elements of the effective force vector $\{f^*(t)\}$ are

$$f_i^* = \{\phi\}_i^T \{f\} \qquad (5.12.4)$$

We find the matrices to be

$$[M^*] = \begin{bmatrix} 106.93 & 0 & 0 \\ 0 & 100.00 & 0 \\ 0 & 0 & 152.00 \end{bmatrix} \text{kips/g} \qquad (5.12.5)$$

$$[K^*] = \begin{bmatrix} 33.56 & 0 & 0 \\ 0 & 150.00 & 0 \\ 0 & 0 & 484.30 \end{bmatrix} \text{kips/in} \qquad (5.12.6)$$

Both $[M^*]$ and $[K^*]$ are diagonal matrices because of the orthogonality of the modes. The matrix product $[M^*]^{-1}[K^*]$ would be a diagonal matrix of ω^2's; thus,

$$[M^*]^{-1}[K^*] = \begin{bmatrix} \omega_1^2 & 0 & 0 \\ 0 & \omega_2^2 & 0 \\ 0 & 0 & \omega_3^2 \end{bmatrix}$$

$$= \begin{bmatrix} 121.18 & 0 & 0 \\ 0 & 579.13 & 0 \\ 0 & 0 & 1230.1 \end{bmatrix} \sec^{-2} \qquad (5.12.7)$$

5.13 SEISMIC INPUT

The displacement of a multidegree system relative to the base is $\{u\}$ and the total displacement is $\{u\} + \{1\}u_g$, where u_g is the ground displacement, a scalar function of time, and $\{1\}$ is a vector with 1's for its elements. The differential equation of motion for response to ground motion is

$$[M]\{\{\ddot{u}\} + \{1\}\ddot{u}_g\} + [K]\{u\} = 0 \qquad (5.13.1)$$

or

$$[M]\{\ddot{u}\} + [K]\{u\} = -[M]\{1\}\ddot{u}_g(t) \qquad (5.13.2)$$

We transform Eq. (5.13.2) into modal coordinates by means of Eq. (5.10.1) and premultiply by $\{\phi\}_p^T$ to get

$$\{\phi\}_p^T[M]\{\phi\}_p\ddot{q}_p + \{\phi\}_p^T[K]\{\phi\}_p q_p = -\{\phi\}_p^T[M]\{1\}\ddot{u}_g(t) \qquad (5.13.3)$$

or

$$\ddot{q}_p + \omega_p^2 q_p = -\frac{\{\phi\}_p^T[M]\{1\}}{\{\phi\}_p^T[M]\{\phi\}_p}\ddot{u}_g(t) \qquad p = 1, 2, \ldots, n \qquad (5.13.4)$$

We define the modal participation factor Γ_p to be

$$\Gamma_p = \frac{\{\phi\}_p^T[M]\{1\}}{\{\phi\}_p^T[M]\{\phi\}_p} \qquad (5.13.5)$$

The modal equations of motion, Eq. (5.13.4), are then

$$\ddot{q}_p + \omega_p^2 q_p = -\Gamma_p\ddot{u}_g(t) \qquad p = 1, 2, \ldots, n \qquad (5.13.6)$$

For our numerical example, the modal participation factors are

$$\Gamma_1 = 1.4028 \qquad \Gamma_2 = -0.5000 \qquad \Gamma_3 = 0.3097 \qquad (5.13.7)$$

and Eq. (5.13.4) for each mode of response to earthquake input is

$$\ddot{q}_1 + (121.18 \text{ sec}^{-2})q_1 = -1.4028\ddot{u}_g(t)$$
$$\ddot{q}_2 + (579.13 \text{ sec}^{-2})q_2 = 0.5000\ddot{u}_g(t) \qquad (5.13.8)$$
$$\ddot{q}_3 + (1230.1 \text{ sec}^{-2})q_3 = -0.3097\ddot{u}_g(t)$$

The equations are completely uncoupled and are amenable to any of the solution techniques developed for SDF systems, including response spectrum techniques.

5.14 EFFECTS OF DAMPING

The equations of motion including damping are

$$[M]\{\ddot{u}\} + [C]\{\dot{u}\} + [K]\{u\} = \{f(t)\} \qquad (5.14.1)$$

Equation (5.10.1) transforms Eq. (5.14.1) into modal coordinates:

$$[M]\{\phi\}_1\ddot{q}_1 + [M]\{\phi\}_2\ddot{q}_2 + [M]\{\phi\}_3\ddot{q}_3$$
$$+ [C]\{\phi\}_1\dot{q}_1 + [C]\{\phi\}_2\dot{q}_2 + [C]\{\phi\}_3\dot{q}_3 \qquad (5.14.2)$$
$$+ [K]\{\phi\}_1 q_1 + [K]\{\phi\}_2 q_2 + [K]\{\phi\}_3 q_3 = \{f\}$$

Equation (5.14.2) is the same as Eq. (5.11.2) except for the insertion of damping terms. To get the first-mode equation, we premultiply each term by $\{\phi\}_1^T$ as we did in Sec. 5.11, obtaining

$$\{\phi\}_1^T[M]\{\phi\}_1\ddot{q}_1 + \{\phi\}_1^T[M]\{\phi\}_2\ddot{q}_2 + \{\phi\}_1^T[M]\{\phi\}_3\ddot{q}_3$$
$$+ \{\phi\}_1^T[C]\{\phi\}_1\dot{q}_1 + \{\phi\}_1^T[C]\{\phi\}_2\dot{q}_2 + \{\phi\}_1^T[C]\{\phi\}_3\dot{q}_3 \qquad (5.14.3)$$
$$+ \{\phi\}_1^T[K]\{\phi\}_1 q_1 + \{\phi\}_1^T[K]\{\phi\}_2 q_2 + \{\phi\}_1^T[K]\{\phi\}_3 q_3 = \{\phi\}_1^T\{f\}$$

Equation (5.14.3) is the same as Eq. (5.11.3) with additional terms due to damping. The coefficients of $\ddot{q}_2$, $\ddot{q}_3$, q_2, and q_3 in Eq. (5.14.3) all vanish because of the orthogonality properties of Eq. (5.8.12). Can we say the same for the coefficients of $\dot{q}_2$ and $\dot{q}_3$? Unfortunately, no. Carrying out the calculations, we find

$$\{\phi\}_1^T[C]\{\phi\}_1 = 0.1678 \text{ kip/(in/sec)}$$
$$\{\phi\}_1^T[C]\{\phi\}_2 = 0.1570 \text{ kip/(in/sec)} \qquad (5.14.4)$$
$$\{\phi\}_1^T[C]\{\phi\}_3 = 0$$

We can express damping in terms of an effective damping matrix $[C^*]$, as we did for effective mass $[M^*]$ and effective stiffness $[K^*]$ in Eq. (5.12.1), the elements of $[C^*]$ being

$$c_{ij}^* = \{\phi\}_i^T[C]\{\phi\}_j \qquad (5.14.5)$$

The resulting damping matrix is

$$[C^*] = \begin{bmatrix} 0.1678 & 0.1570 & 0 \\ 0.1570 & 1.2500 & 0.5000 \\ 0 & 0.5000 & 2.4216 \end{bmatrix} \text{kip}/(\text{in}/\text{sec}) \qquad (5.14.6)$$

Whereas $[M^*]$ and $[K^*]$ were both diagonal matrices, $[C^*]$ is not, and the equations remain coupled in the velocity terms. If the damping matrix were some multiple of the inertia matrix, or some multiple of the stiffness matrix, or some linear combination of the two, then the effective damping matrix $[C^*]$ would be diagonal and the velocity coupling would vanish. Sometimes either mass-proportional damping or stiffness-proportional damping is assumed to exist in order to circumvent the velocity-coupling problem. Mass-proportional damping is a bit easier to handle computationally than stiffness-proportional damping, but it leads to a lesser fraction of critical damping for higher modes than for the fundamental mode. The reverse is true for stiffness-proportional damping, which might be more palatable to the intuition because the higher modes tend to decay more rapidly than the first.

The damping matrix $[C]$ is very difficult to determine, either theoretically or experimentally, and the velocity-coupling problem is usually eluded either by ignoring it, that is, by taking the off-diagonal terms of $[C^*]$ to be zero, whether they are or not, or else by temporarily setting aside damping, finding the modes for the undamped system, and then inserting a fraction of critical damping into each modal equation.

There is a way of determining normal modes that yields completely uncoupled equations for a damped multidegree-of-freedom system. Essentially, the procedure is to recast the N second-order differential equations of motion into the form of $2N$ first-order equations and then to determine the eigenvalues and eigenvectors of the resulting $2N \times 2N$ dynamic matrix. The eigenvalues and eigenvectors turn out to be complex conjugate pairs. For a stable dynamic system, the eigenvalues have negative real parts, which give the fraction of critical damping for the particular mode, and the imaginary parts give the frequency. The procedure is mathematically elegant but not particularly meaningful in view of the inherent uncertainty in the damping matrix $[C]$ in the first place.

5.15 NUMERICAL EXAMPLE: RESPONSE TO DRIVING FORCE

We now return to the example structure of Fig. 5.5 with the driving force of Fig. 5.7, a single sine pulse at the second mass. We solved the problem numerically in Sec. 5.5; here we solve it by normal modes. The undamped equations are

$$\ddot{q}_p + \omega_p^2 q_p = \frac{\{\phi\}_p^T\{f\}}{\{\phi\}_p^T[M]\{\phi\}_p} \qquad p = 1, 2, 3 \qquad (5.15.1)$$

Now let us insert damping for each mode, but ignore the velocity coupling, that is, neglect the off-diagonal terms of $[C^*]$. We then get

$$\ddot{q}_p + \frac{\{\phi\}_p^T[C]\{\phi\}_p}{\{\phi\}_p^T[M]\{\phi\}_p}\dot{q}_p + \omega_p^2 q_p = \frac{\{\phi\}_p^T\{f\}}{\{\phi\}_p^T[M]\{\phi\}_p} \qquad (5.15.2)$$

or

$$\ddot{q}_p + 2\zeta_p \omega_p \dot{q}_p + \omega_p^2 q_p = f_p^*(t) \qquad (5.15.3)$$

where

$$2\zeta_p \omega_p = \frac{\{\phi\}_p^T[C]\{\phi\}_p}{\{\phi\}_p^T[M]\{\phi\}_p} \qquad (5.15.4)$$

and

$$f_p^*(t) = \frac{\{\phi\}_p^T\{f\}}{\{\phi\}_p^T[M]\{\phi\}_p} \qquad (5.15.5)$$

Both f_1 and f_3 were zero in our example. Evaluation of the terms gives the three equations of motion:

$$\ddot{q}_1 + 0.6059\dot{q}_1 + 121.18q_1 = 123.87f_2(t)$$
$$\ddot{q}_2 + 4.8261\dot{q}_2 + 579.13q_2 = -96.522f_2(t) \qquad (5.15.6)$$
$$\ddot{q}_3 + 6.1507\dot{q}_3 + 1230.1q_3 = -87.139f_2(t)$$

The fractions of critical damping turn out to be

$$\zeta_1 = 0.0275 \qquad \zeta_2 = 0.1003 \qquad \zeta_3 = 0.0877 \qquad (5.15.7)$$

Now we take the same force as before,

$$f_2 = P \sin(\pi t/t_1) \qquad 0 \le t < t_1$$
$$= 0 \qquad t \ge t_1 \qquad (5.15.8)$$

where $P = 150$ kips and $t_1 = 0.2$ sec. Solving the equations either analytically or numerically, we get the three modal displacement–time curves shown in Fig. 5.12.

The transformation of Eq. (5.10.1) gives the same results in u coordinates, as shown in Fig. 5.13. The curves of Fig. 5.13 are very close to those of Fig. 5.8, but not quite identical, for in the numerical solution shown in Fig. 5.8, we included all the effects of damping, whereas here we have ignored the off-diagonal terms of $[C^*]$.

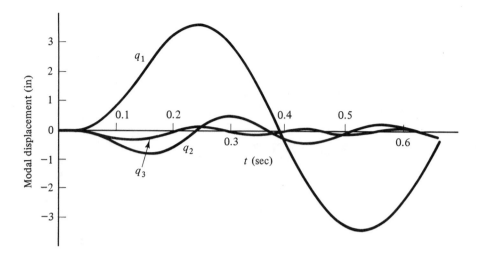

Figure 5.12 Displacements in modal coordinates.

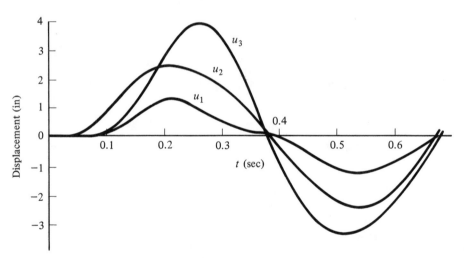

Figure 5.13 Displacements in u coordinates.

5.16 NUMERICAL EXAMPLE: RESPONSE TO EARTHQUAKE

The modal equations for response to earthquake, with damping inserted, are

$$\ddot{q}_p + 2\zeta_p\omega_p\dot{q}_p + \omega_p^2 q_p = -\Gamma_p\ddot{u}_g(t) \qquad (5.16.1)$$

We could solve these numerically to get q_1, q_2, and q_3 as functions of time, and then invoke Eq. (5.10.1) to get u_1, u_2, and u_3 as functions of time. Then from the u's, we could compute any other response variables of interest, such as bending moments in frame members or axial forces in columns. Alternatively, we may employ response spectrum techniques, for each modal equation is the same as the equation for a SDF system except for the multiplier Γ_p on the ground acceleration. Thus,

$$|q_p|_{\max} = |\Gamma_p| \ \text{SD} \qquad (5.16.2)$$

A difficulty arises when we try to convert the q's back into u coordinates, for although we know the maximum magnitude of each q_p, we do not know when it occurs or in which direction. In employing response spectra, we forego any information about the time variation of the response. Thus, while we can determine the maximum displacements, or for that matter the maximum value of any other response parameter, for each mode, we can no longer combine modes to get the total response. The loss is not as tragic as it might first seem, however. We can establish an upper bound on each response parameter as the sum of the absolute values of the modal responses. Sometimes the sum of the absolute values of the first two or three modal responses is accepted as an estimate of the total response. More frequently, the total response is taken to be the square root of the sum of the squares of the modal responses, sometimes called the RSS response. *Caution:* the particular response variable sought must be evaluated *for each mode* first, *before* combining modal responses. For example, to get the maximum drift within the second story, compute $|u_3 - u_2|_{\max}$ for each mode first and then combine them. *Do not* combine the modal u_3's to get $|u_3|_{\max}$; combine modal u_2's to get $|u_2|_{\max}$, and then take the difference to get the story drift.

Consider the example structure of Fig. 5.5 subjected to the N11°W component of the Eureka, California, earthquake of December 21, 1954. The response spectra for that earthquake component are shown in Fig. 4.3. For the three modes and combined responses, we get the results shown in Table 5.3.

TABLE 5.3   Modal Response to Earthquake

Mode number, p	1	2	3	Sum	RSS
Period T_p (sec)	0.571	0.261	0.179	—	—
Damping ζ_p	0.028	0.100	0.088	—	—
Γ_p	1.403	−0.500	0.310	—	—
SD (in)	1.36	0.25	0.09	—	—
Max. $|q_p|$ (in)	1.91	0.123	0.028	—	—
Max. $|u_1|$ (in)	0.600	0.062	0.028	0.690	0.604
Max. $|u_2|$ (in)	1.313	0.062	0.019	1.393	1.314
Max. $|u_3|$ (in)	1.913	0.123	0.009	2.045	1.917
Max. $|u_2 - u_1|$ (in)	0.712	0	0.047	0.759	0.714
Max. $|u_3 - u_2|$ (in)	0.600	0.185	0.028	0.813	0.629

5.17 BASE-SHEAR EQUIVALENT MASS

The response of mode p to earthquake motion is given by

$$\ddot{q}_p + 2\zeta_p\omega_p\dot{q}_p + \omega_p^2 q_p = -\Gamma_p\ddot{u}_g(t) \qquad (5.16.1)$$

Equation (5.9.1) converts this q_p to its equivalent in u coordinates:

$$\{u(t)\} = \{\phi\}_p q_p(t) \qquad (5.17.1)$$

The restoring forces at the floor levels are

$$\{R\} = [K]\{u\} = [K]\{\phi\}_p q_p(t) = \omega_p^2[M]\{\phi\}_p q_p(t) \qquad (5.17.2)$$

and the base shear, the total force transmitted from the ground to the structure, is the sum of the restoring forces:

$$V(t) = \{1\}^T\{R(t)\} = \omega_p^2(\{1\}^T[M]\{\phi\}_p)q_p(t) \qquad (5.17.3)$$

For comparison, the equation of motion for a linear SDF system driven by an earthquake is

$$\ddot{u} + 2\zeta\omega_n\dot{u} + \omega_n^2 u = -\ddot{u}_g(t) \qquad (2.14.1)$$

and the base shear is

$$V(t) = ku(t) = \omega_n^2 mu(t) \qquad (5.17.4)$$

The modal displacement q_p at any instant is Γ_p times the displacement u of the SDF system. Thus, if the SDF system has the same frequency and fraction of critical damping as mode p of the MDF system, and if its mass is

$$m = m_p^v = \{1\}^T[M]\{\phi\}_p\Gamma_p = \frac{(\{\phi\}_p^T[M]\{1\})^2}{\{\phi\}_p^T[M]\{\phi\}_p} \qquad (5.17.5)$$

then the base shear at any instant will be the same for the SDF system as for mode p of the MDF system. This mass is called the **base-shear equivalent mass** for mode p. The concept is useful in earthquake engineering, for it facilitates the adaptation of response spectrum techniques to MDF systems. The maximum base shear for any mode of response to an earthquake is simply the product of the base-shear equivalent mass and the spectral acceleration for that frequency and fraction of critical damping:

$$V_p = m_p^v\,(\text{PSA})_p \qquad (5.17.6)$$

If $[M]$ is diagonal, then

$$m^v = \left(\sum m\phi\right)^2 \Big/ \sum m\phi^2 \qquad (5.17.7)$$

The sum of the base-shear equivalent masses for all modes is equal to the total mass in the system.

For the system of Fig. 5.3, the values are

$$m_1^v = 210.42 \text{ kips/g}$$
$$m_2^v = 25.00 \text{ kips/g}$$
$$\underline{m_3^v = 14.58 \text{ kips/g}}$$
$$\sum_p m_p^v = 250.00 \text{ kips/g}$$

(5.17.8)

PROBLEMS

5.1 The double pendulum in the figure consists of two 10-lb weights, which may be taken as point masses, suspended by weightless inextensible rods 30 in long. Displacements are small and in the plane of the paper. There is no damping.

Find the two natural frequencies, two natural periods, and two mode shapes.

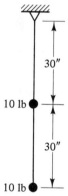

5.2 The pendulum of Problem 5.1 is displaced to the initial configuration shown in the figure and released with zero initial velocity. Find the equations of free vibration as functions of time in modal coordinates and in u coordinates.

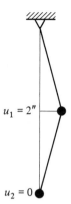

$u_1 = 2''$

$u_2 = 0$

5.3 A weightless undamped cantilever beam of length $2l$ and stiffness EI has a point mass m attached at the tip and a point mass $2m$ attached at midlength, as shown in the figure. It is subjected to a transverse driving force $f(t)$ at the midpoint.

Find the frequencies, the mode shapes, and the modal differential equations of motion.

$f(t)$

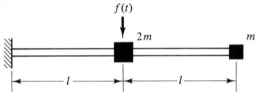

$2m$ m

$\leftarrow\!\!\!-\!\!\!-\!\!\!- l \!\!\!-\!\!\!-\!\!\!-\!\!\!\rightarrow\!\!\!\leftarrow\!\!\!-\!\!\!-\!\!\!- l \!\!\!-\!\!\!-\!\!\!-\!\!\!\rightarrow$

5.4 The undamped triple pendulum shown in the figure has three point masses, $3m$, $2m$, and m, suspended on weightless inextensible rods of length l. Find the three natural frequencies and mode shapes.

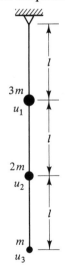

$3m$
u_1

$2m$
u_2

m
u_3

5.5 The suspension point of the pendulum of Problem 5.4 moves laterally, as shown in the figure, in a sinusoidal motion of amplitude A and frequency ω, that is,

$$u_g = A \sin \omega t$$

Find the three modal differential equations of motion.

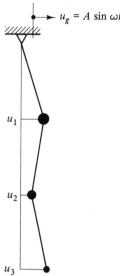

5.6 The figure shows one bent of a three-story single-bay structure. Treat this as a three-degree-of-freedom system, an ideal weightless frame with flexible columns and rigid girders, with its mass concentrated at the floor levels. Neglect damping, lengthening and shortening of frame members, and the effects of gravity forces on column bending moments and column stiffness.

A lateral driving force $f(t)$ acts at the middle floor.

(a) Find the inertia and stiffness matrices $[M]$ and $[K]$.

(b) Find the frequency equation and the three frequencies and mode shapes.

(c) Write the three modal differential equations of motion.

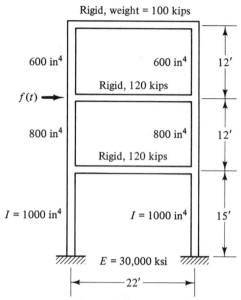

Rigid, weight = 100 kips

600 in^4 600 in^4 12'

Rigid, 120 kips

f(t) →

800 in^4 800 in^4 12'

Rigid, 120 kips

I = 1000 in^4 I = 1000 in^4 15'

E = 30,000 ksi

22'

5.7 The structure shown in the figure is identical with that of Problem 5.6 except that the girders are flexible and their flexibility is to be taken into account. The driving force is the same.
   (a) Find the inertia and stiffness matrices $[M]$ and $[K]$.
   (b) Find the frequency equation and the three frequencies and mode shapes.
   (c) Write the three modal differential equations of motion.

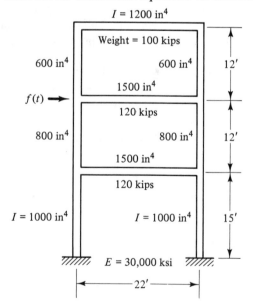

I = 1200 in^4

Weight = 100 kips

600 in^4 600 in^4 12'

1500 in^4

f(t) →

120 kips

800 in^4 800 in^4 12'

1500 in^4

120 kips

I = 1000 in^4 I = 1000 in^4 15'

E = 30,000 ksi

22'

CHAPTER SIX

Finding Normal Modes of Multidegree Systems

6.1 INTRODUCTION

In developing the normal modes procedure in Chapter 5, we derived the equation

$$[[K] - \omega^2[M]]\{\phi\} = 0 \qquad (5.6.11)$$

which was valid for any mode. This led to the frequency equation

$$|[K] - \omega^2[M]| = 0 \qquad (5.6.12)$$

We expanded the determinant of Eq. (5.6.12) into a polynomial in ω^2, and found that there was a modal vector $\{\phi\}$ associated with each root ω^2. The procedure was mathematically correct but not entirely practical for systems with more than a few degrees of freedom, because expanding a determinant into a polynomial and finding its roots is a laborious process. It is better to seek other ways. We shall consider several ways in this chapter.

6.2 STODOLA PROCESS FOR FUNDAMENTAL MODE

The Stodola process is among the more attractive alternatives for finding modes. For any mode, say mode p, Eq. (5.6.11) can be written as

$$[K]\{\phi\}_p = \omega_p^2[M]\{\phi\}_p \qquad (6.2.1)$$

Now premultiply both sides of Eq. (6.2.1) by $[K]^{-1}/\omega_p^2$ and define a new matrix:

$$[D] = [K]^{-1}[M] = \text{the dynamic flexibility matrix}$$

The result is

$$\{\phi\}_p/\omega_p^2 = [D]\{\phi\}_p \qquad (6.2.2)$$

In the Stodola process, we use Eq. (6.2.2) iteratively to determine the fundamental mode. We start with an arbitrary vector $\{U\}_1$, premultiply it by the dynamic flexibility matrix $[D]$ to get a new vector $\{V\}_1$, and divide it by its largest element N_1 to get $\{U\}_2$. Unless $\{U\}_1$ was exactly orthogonal to the first mode vector $\{\phi\}_1$, $\{U\}_2$ will be more nearly proportional to $\{\phi\}_1$ than was $\{U\}_1$. The reasons will be examined in Sec. 6.4. Repeating the process brings us closer and closer to the fundamental mode.

In more formal notation, the iteration is

$$\{V\}_j = [D]\{U\}_j$$
$$N_j = \text{largest element of } \{V\}_j \qquad j = 1, 2, 3, \ldots \qquad (6.2.3)$$
$$\{U\}_{j+1} = \{V\}_j/N_j$$

As j increases, $\{U\}_j$ approaches $\{\phi\}_1$ and N_j approaches $1/\omega_1^2$.

For example, consider the five-story shear building of Fig. 6.1, with floor weights and story stiffnesses as shown. Take the floors to be infinitely rigid compared with the columns. The inertia and stiffness matrices are then

$$[M] = \begin{bmatrix} 140 & 0 & 0 & 0 & 0 \\ 0 & 120 & 0 & 0 & 0 \\ 0 & 0 & 120 & 0 & 0 \\ 0 & 0 & 0 & 120 & 0 \\ 0 & 0 & 0 & 0 & 100 \end{bmatrix} \text{kip/g} \qquad (6.2.4)$$

$$[K] = \begin{bmatrix} 800 & -400 & 0 & 0 & 0 \\ -400 & 600 & -200 & 0 & 0 \\ 0 & -200 & 400 & -200 & 0 \\ 0 & 0 & -200 & 300 & -100 \\ 0 & 0 & 0 & -100 & 100 \end{bmatrix} \text{kip/in} \qquad (6.2.5)$$

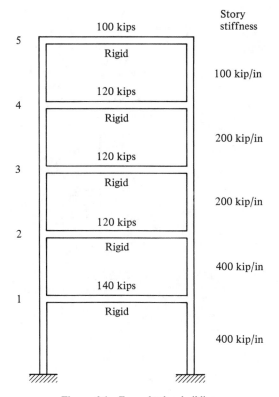

Figure 6.1 Example shearbuilding.

The static flexibility matrix, the inverse of the stiffness matrix, is

$$[K]^{-1} = \begin{bmatrix} 0.0025 & 0.0025 & 0.0025 & 0.0025 & 0.0025 \\ 0.0025 & 0.0050 & 0.0050 & 0.0050 & 0.0050 \\ 0.0025 & 0.0050 & 0.0100 & 0.0100 & 0.0100 \\ 0.0025 & 0.0050 & 0.0100 & 0.0150 & 0.0150 \\ 0.0025 & 0.0050 & 0.0100 & 0.0150 & 0.0250 \end{bmatrix} \text{ in/kip} \qquad (6.2.6)$$

and the dynamic flexibility matrix is

$$[D] = [K]^{-1}[M] = \begin{bmatrix} 0.35 & 0.30 & 0.30 & 0.30 & 0.25 \\ 0.35 & 0.60 & 0.60 & 0.60 & 0.50 \\ 0.35 & 0.60 & 1.20 & 1.20 & 1.00 \\ 0.35 & 0.60 & 1.20 & 1.80 & 1.50 \\ 0.35 & 0.60 & 1.20 & 1.80 & 2.50 \end{bmatrix} \text{ in/g} \qquad (6.2.7)$$

We take the initial vector $\{U\}_1$ to be $\{1\}$, which is as convenient as any, and perform the iteration of Eq. (6.2.3). Table 6.1 shows the results. Convergence is fairly rapid.

TABLE 6.1 Stodola Iteration for Fundamental Mode

j	1	2	3	4	5	6
$\{U\}_j$ (1)	$\begin{Bmatrix} 1 \\ 1 \\ 1 \\ 1 \\ 1 \end{Bmatrix}$	$\begin{Bmatrix} 0.2326 \\ 0.4108 \\ 0.6744 \\ 0.8450 \\ 1.0000 \end{Bmatrix}$	$\begin{Bmatrix} 0.1765 \\ 0.3372 \\ 0.6109 \\ 0.8061 \\ 1.0000 \end{Bmatrix}$	$\begin{Bmatrix} 0.1694 \\ 0.3262 \\ 0.5991 \\ 0.7979 \\ 1.0000 \end{Bmatrix}$	$\begin{Bmatrix} 0.1683 \\ 0.3245 \\ 0.5970 \\ 0.7963 \\ 1.0000 \end{Bmatrix}$	$\begin{Bmatrix} 0.1681 \\ 0.3242 \\ 0.5967 \\ 0.7961 \\ 1.0000 \end{Bmatrix}$
$\{V\}_j$ (in/g)	$\begin{Bmatrix} 1.5000 \\ 2.6500 \\ 4.3500 \\ 5.4500 \\ 6.4500 \end{Bmatrix}$	$\begin{Bmatrix} 0.9105 \\ 1.7395 \\ 3.1512 \\ 4.1581 \\ 5.1581 \end{Bmatrix}$	$\begin{Bmatrix} 0.8381 \\ 1.6144 \\ 2.9646 \\ 3.9482 \\ 4.9482 \end{Bmatrix}$	$\begin{Bmatrix} 0.8263 \\ 1.5932 \\ 2.9314 \\ 3.9102 \\ 4.9102 \end{Bmatrix}$	$\begin{Bmatrix} 0.8242 \\ 1.5896 \\ 2.9256 \\ 3.9034 \\ 4.9034 \end{Bmatrix}$	$\begin{Bmatrix} 0.8239 \\ 1.5890 \\ 2.9246 \\ 3.9022 \\ 4.9022 \end{Bmatrix}$
ω_1 from N_j (rad/sec)	7.7368	8.6516	8.8332	8.8673	8.8735	8.8746
ω_1 from Rayleigh (rad/sec)	9.0176	8.8782	8.8749	8.8748	8.8748	8.8748

The true fundamental mode and frequency are

$$\{\phi\}_1 = \begin{Bmatrix} 0.16806 \\ 0.32412 \\ 0.59657 \\ 0.79600 \\ 1.00000 \end{Bmatrix} \tag{6.2.8}$$

$$\omega_1 = 8.87478 \text{ rad/sec}$$

After six cycles, vector $\{U\}$ and the largest element N give the fundamental mode and frequency to an accuracy of 10^{-4}.

6.3 THE RAYLEIGH QUOTIENT

The last line of Table 6.1 gives the frequency computed at the end of each cycle from the **Rayleigh quotient,** which we now will define.

From Eq. (5.6.11), we have, for mode p of free vibration.

$$[K]\{\phi\}_p = \omega_p^2[M]\{\phi\}_p \tag{6.3.1}$$

We premultiply both sides of Eq. (6.3.1) by $\{\phi\}_p^T$ to get

$$\{\phi\}_p^T[K]\{\phi\}_p = \omega_p^2\{\phi\}_p^T[M]\{\phi\}_p \tag{6.3.2}$$

or, upon dividing by $\{\phi\}_p^T[M]\{\phi\}_p$,

$$\omega_p^2 = \frac{\{\phi\}_p^T[K]\{\phi\}_p}{\{\phi\}_p^T[M]\{\phi\}_p} \tag{6.3.3}$$

The right side of Eq. (6.3.3) is the Rayleigh quotient.

We could also derive Eq. (6.3.3) from energy considerations as follows: if the system were oscillating in free vibration in mode p, its displacement and velocity vectors would be

$$\{u\} = \{\phi\}_p A \sin \omega_p t \tag{6.3.4}$$

and

$$\{\dot{u}\} = \{\phi\}_p \omega_p A \cos \omega_p t \tag{6.3.5}$$

At the extreme displacement, the velocity, and therefore the kinetic energy, is zero; the total energy in the system is the strain energy:

$$E = \frac{1}{2}\{u\}^T[K]\{u\} = (A^2/2)\{\phi\}_p^T[K]\{\phi\}_p \tag{6.3.6}$$

At the extreme velocity, the displacement, and therefore the strain energy, is zero; the total energy in the system is the kinetic energy:

$$E = \frac{1}{2}\{\dot{u}\}^T[M]\{\dot{u}\} = (A^2/2)\omega_p^2\{\phi\}_p^T[M]\{\phi\}_p \tag{6.3.7}$$

Equating these two expressions for energy in the system yields Eq. (6.3.2), from which we obtained the Rayleigh quotient in Eq. (6.3.3).

If the mode shape is known only approximately, the Rayleigh quotient still gives a close estimate of the frequency. In the Stodola process, we can take the last computed vector $\{V\}$ as an approximate mode shape and compute the frequency from

$$\omega_1^2 = \frac{\{V\}^T[K]\{V\}}{\{V\}^T[M]\{V\}} \tag{6.3.8}$$

But in the iteration, we had

$$\{V\} = [D]\{U\} = [K]^{-1}[M]\{U\} \tag{6.3.9}$$

Thus,

$$[K]\{V\} = [M]\{U\} \tag{6.3.10}$$

which, substituted in Eq. (6.3.8), gives us

$$\omega_1^2 = \frac{\{V\}^T[M]\{U\}}{\{V\}^T[M]\{V\}} \tag{6.3.11}$$

If $\{V\}$ were the exact mode, the frequency would, of course, be identical with that obtained from N, the largest element of $\{V\}$. If convergence is incomplete, the Rayleigh quotient gives the better approximation. Any error in the first-mode frequency computed by the Rayleigh quotient is always on the high side. We see in Table 6.1 that the Rayleigh quotient gives a frequency within 2 percent of the true frequency in the first cycle, and a frequency more accurate after three cycles than N gives after six.

6.4 STODOLA CONVERGENCE

In the foregoing example, the Stodola process converged to the first mode. We now explore the reasons.

We began the iteration with an initial vector $\{U\}$ that was arbitrary. Whatever the choice of $\{U\}$, it is some linear combination of unknown modes:

$$\{U\} = \{\phi\}_1 q_1 + \{\phi\}_2 q_2 + \{\phi\}_3 q_3 + \dots \tag{6.4.1}$$

where the q's are undetermined constants. We then iterate according to Eq. (6.2.3), computing

$$\{V\} = [D]\{U\} = [D]\{\phi\}_1 q_1 + [D]\{\phi\}_2 q_2 + [D]\{\phi\}_3 q_3 + \dots \tag{6.4.2}$$

But for every one of the modes, all of which are yet unknown

$$[D]\{\phi\}_p = \{\phi\}_p / \omega_p^2 \tag{6.4.3}$$

Thus,

$$\{V\} = \{\phi\}_1 q_1 / \omega_1^2 + \{\phi\}_2 q_2 / \omega_2^2 + \{\phi\}_3 q_3 / \omega_3^2 + \dots \tag{6.4.4}$$

Comparing Eq. (6.4.4) with Eq. (6.4.1), we see that every modal coefficient q_p on the right side of Eq. (6.4.1) has been divided by its corresponding ω_p^2. If the first-mode frequency is distinct, then $\omega_1^2 < \omega_2^2 \leq \omega_3^2 \dots$ and the first-mode component is magnified relative to the other components. If the first-mode frequency is distinct and the first-mode component is nonzero, no matter how small, the iteration of Eq. (6.2.3) converges to the first mode. Convergence is quicker, of course, if the initial vector $\{U\}_1$ is close to the first mode. For hand calculation, there is some merit in seeking a close initial vector. With the computer, it is simpler to take $\{U\}_1 = \{1\}$, which always has a nonzero first-mode component. The cost in extra computing is perhaps one cycle of iteration.

The iteration of Table 6.1 began with $\{U\}_1 = \{1\}$. The vector $\{V\}_1$ computed in the first cycle is then the static deflected shape for a lateral force equal to the weight of the structure. This shape is often used as an approximation to the first-mode shape, and the Rayleigh quotient at this stage gives a frequency often called the Rayleigh frequency.

6.5 STODOLA PROCESS FOR THE SECOND MODE

In each cycle of the Stodola process, each modal component of $\{U\}$ is divided by its frequency ω^2. If we could choose the initial $\{U\}$ to have a zero first-mode component, and if the second-mode frequency were distinct, then the second mode would get the greatest magnification and the process ought to converge to the second mode.

We can make the first-mode component zero in the following manner. Any vector $\{U\}$ is a linear combination of all modes:

$$\{U\} = \{\phi\}_1 q_1 + \{\phi\}_2 q_2 + \{\phi\}_3 q_3 + \ldots \tag{6.4.1}$$

Knowing $\{\phi\}_1$, we premultiply the terms of Eq. (6.4.1) by $\{\phi\}_1^T[M]$ to get

$$\{\phi\}_1^T[M]\{U\} = \{\phi\}_1^T[M]\{\phi\}_1 q_1 + \{\phi\}_1^T[M]\{\phi\}_2 q_2$$
$$+ \{\phi\}_1^T[M]\{\phi\}_3 q_3 + \ldots \tag{6.5.1}$$

By virtue of modal orthogonality, all terms on the right side of Eq. (6.5.1) are zero except the first; thus,

$$q_1 = \frac{\{\phi\}_1^T[M]\{U\}}{\{\phi\}_1^T[M]\{\phi\}_1} \tag{6.5.2}$$

and the vector $\{\{U\} - \{\phi\}_1 q_1\}$ has a zero first-mode component.

In theory, we could take an arbitrary vector $\{U\}$, remove the first-mode component, use the remainder as the initial vector, and the Stodola iteration of Eq. (6.2.3) ought then to converge to the second mode. In practice, it doesn't work quite that well. Ordinarily, we will not know $\{\phi\}_1$ precisely, for our calculations will not have been executed with infinite precision. Thus, the starting vector, instead of being exactly orthogonal to the exact first mode, will be approximately orthogonal to an approximate first mode. Removal of the first mode is usually called "sweeping it out," an apt description. The mode is swept out, not scoured out, and the remainder is not sterile — it still contains a little first-mode residue. That minute first-mode component would be magnified with each cycle of the iteration, and would ultimately dominate. The iteration would again converge to the first mode.

The remedy is to sweep out the first mode in each cycle. We have

$$\{U\} - \{\phi\}_1 q_1 = \{U\} - \{\phi\}_1 \frac{\{\phi\}_1^T[M]\{U\}}{\{\phi\}_1^T[M]\{\phi\}_1}$$

$$= \left[[I] - \frac{\{\phi\}_1\{\phi\}_1^T[M]}{\{\phi\}_1^T[M]\{\phi\}_1}\right]\{U\} \tag{6.5.3}$$

Define the sweeping matrix $[S]_1$ to be

$$[S]_1 = \left[[I] - \frac{\{\phi\}_1\{\phi\}_1^T[M]}{\{\phi\}_1^T[M]\{\phi\}_1}\right] \tag{6.5.4}$$

Then premultiplying any arbitrary vector $\{U\}$ by the sweeping matrix $[S]_1$ removes the first-mode component.

Now we define a new dynamic flexibility matrix:

$$[D]_2 = [D][S]_1 \tag{6.5.5}$$

Premultiplying vector $\{U\}$ by the dynamic matrix $[D]_2$ in a single operation sweeps out its first-mode component and premultiplies the remainder by the dynamic flexibility matrix $[D]$. We iterate with the modified dynamic matrix just as before:

$$\{V\}_j = [D]_2\{U\}_j$$
$$N_j = \text{largest element of } \{V\}_j \qquad j = 1, 2, 3, \ldots \tag{6.5.6}$$
$$\{U\}_{j+1} = \{V\}_j/N_j$$

This time, as j increases, $\{U\}_j$ approaches the second mode $\{\phi\}_2$ and N_j approaches $1/\omega_2^2$. Again the Rayleigh quotient gives a better estimate of the frequency:

$$\omega_2^2 = \frac{\{V\}^T[M]\{U\}}{\{V\}^T[M]\{V\}} \tag{6.5.7}$$

For the example of Fig. 6.1, for which the first-mode vector was given in Eq. (6.2.8), we obtain

$$[S]_1 = \left[[I] - \frac{\{\phi\}_1\{\phi\}_1^T[M]}{\{\phi\}_1^T[M]\{\phi\}_1}\right]$$

$$= \begin{bmatrix} 0.98320 & -0.02778 & -0.05113 & -0.06822 & -0.07142 \\ -0.03241 & 0.94642 & -0.09681 & -0.13158 & -0.13775 \\ -0.05965 & -0.09861 & 0.81850 & -0.24218 & -0.25353 \\ -0.07959 & -0.13518 & -0.24218 & 0.67687 & -0.33829 \\ -0.09999 & -0.16530 & -0.30424 & -0.40595 & 0.57501 \end{bmatrix}$$
$$\tag{6.5.8}$$

$$[D]_2 = [D][S]_1$$

$$= \begin{bmatrix} 0.26762 & 0.16382 & 0.04936 & -0.03443 & -0.10012 \\ 0.19113 & 0.33737 & 0.11661 & -0.04498 & -0.17523 \\ 0.05758 & 0.11661 & 0.31029 & 0.01286 & -0.24282 \\ -0.04017 & -0.04498 & 0.01286 & 0.21601 & -0.15828 \\ -0.14016 & -0.21028 & -0.29138 & -0.18994 & 0.41673 \end{bmatrix}$$

$$(6.5.9)$$

Table 6.2 shows the iteration. Convergence is less rapid this time; it takes about twice as many cycles as for the first mode to achieve the same accuracy.

TABLE 6.2   Stodola Iteration for Second Mode

j	1	2	3	4	5	6
$\{U\}_j$ (1)	$\begin{Bmatrix} 1 \\ 1 \\ 1 \\ 1 \\ 1 \end{Bmatrix}$	$\begin{Bmatrix} 0.8149 \\ 1.0000 \\ 0.5990 \\ -0.0343 \\ -0.9768 \end{Bmatrix}$	$\begin{Bmatrix} -0.5674 \\ -0.8178 \\ -0.6516 \\ -0.0858 \\ 1.0000 \end{Bmatrix}$	$\begin{Bmatrix} -0.4748 \\ -0.7225 \\ -0.6566 \\ -0.1437 \\ 1.0000 \end{Bmatrix}$	$\begin{Bmatrix} -0.4369 \\ -0.6791 \\ -0.6559 \\ -0.1712 \\ 1.0000 \end{Bmatrix}$	$\begin{Bmatrix} -0.4201 \\ -0.6592 \\ -0.6548 \\ -0.1843 \\ 1.0000 \end{Bmatrix}$
$\{V\}_j$ (in/g)	$\begin{Bmatrix} 0.3463 \\ 0.4249 \\ 0.2545 \\ -0.0146 \\ -0.4150 \end{Bmatrix}$	$\begin{Bmatrix} 0.5104 \\ 0.7357 \\ 0.5861 \\ 0.0772 \\ -0.8996 \end{Bmatrix}$	$\begin{Bmatrix} -0.4152 \\ -0.6317 \\ -0.5741 \\ -0.1256 \\ 0.8744 \end{Bmatrix}$	$\begin{Bmatrix} -0.3730 \\ -0.5798 \\ -0.5600 \\ -0.1462 \\ 0.8538 \end{Bmatrix}$	$\begin{Bmatrix} -0.3548 \\ -0.5566 \\ -0.5529 \\ -0.1556 \\ 0.8444 \end{Bmatrix}$	$\begin{Bmatrix} -0.3465 \\ -0.5460 \\ -0.5494 \\ -0.1560 \\ 0.8400 \end{Bmatrix}$
ω_2 from N_j (rad/sec)	30.1440	20.7169i*	21.0132	21.2648	21.3831	21.4387
ω_2 from Rayleigh (rad/sec)	22.9719	21.8047	21.5572	21.5031	21.4913	21.4887

*In this cycle, the largest element of $\{V\}$ is negative.

The true second mode and frequency are

$$\{\phi\}_2 = \begin{Bmatrix} -0.40590 \\ -0.64191 \\ -0.65331 \\ -0.19593 \\ 1.00000 \end{Bmatrix}$$

$$(6.5.10)$$

$$\omega_2 = \quad 21.48795 \text{ rad/sec}$$

6.6 EXTENSION TO HIGHER MODES

To get higher modes by the Stodola process, we compute them in sequence, modifying the sweeping matrix for each mode to remove all modes lower than the one sought. The following algorithm will compute any desired number of modes.

For mode 1:

(a) Set

$$[S]_0 = [I] \qquad (6.6.1)$$

(b) Compute

$$[D]_1 = [K]^{-1}[M][S]_0 \qquad (6.6.2)$$

(c) Set

$$\{U\} = \{1\} \qquad (6.6.3)$$

(d) Iteratively compute

$$\{V\} = [D]_1\{U\} \qquad (6.6.4)$$

according to Eq. (6.2.3). The iteration converges to the first mode.

For mode $p + 1$, having modes $1, 2, \ldots, p$:

(a) Compute

$$[S]_p = [S]_{p-1} - \frac{\{\phi\}_p \{\phi\}_p^T [M]}{\{\phi\}_p^T [M]\{\phi\}_p} \qquad (6.6.5)$$

(b) Compute

$$[D]_{p+1} = [K]^{-1}[M][S]_p \qquad (6.6.6)$$

(c) Set

$$\{U\} = \{1\} \qquad (6.6.7)$$

(d) Iteratively compute

$$\{V\} = [D]_{p+1}\{U\} \qquad (6.6.8)$$

as in Eq. (6.2.3). The iteration converges to mode $p + 1$.

Program 6.1 executes the algorithm, and Table 6.3 shows the five modes computed for the example of Fig. 6.1.

Program 6.1: Modes by the Stodola Process

```
1 '  Input:    N = degrees of freedom,
2 '            MODES = No. of modes sought,
3 '            W(1) . . . W(N) = floor weights,
4 '            F(1,1) . . . F(N,N) = flexibility matrix,
5 '
6 '  Output:   Modes, frequencies, periods,
7 '            modal participation factors,
8 '            base shear equivalent weights.

10    GOSUB 1000                  ' Housekeeping instructions
20    GOSUB 2000                  ' Get data, set [S] = [I]
30    FOR MODE=1 TO MODES
40       GOSUB 3000               ' Update [D], set {U} = {1}
50       GOSUB 4000               ' Execute Stodola iteration
60       GOSUB 5000               ' Get mode properties
70       IF MODE<MODES THEN GOSUB 6000    ' Update [S]
80    NEXT MODE
90    GOTO 7000                   ' Print mode properties

1000 DEFINT I-N                        ' Housekeeping
1010 DIM W(10),U(10),V(10),FREQ(10),PERIOD(10)
1020 DIM GAMMA(10),EQUIVWT(10)
1030 DIM F(10,10),S(10,10),D(10,10),PHI(10,10)
1040 F2$=" ####.##       "
1050 F4$=" ##.####       "
1060 F5$=" #.#####       "
1070 E5$=" #.#####^^^^ "
1080 GRAV=386.088
1090 PI=3.141593
1100 RETURN

2000 READ N, MODES                     ' Get data
2010 FOR I=1 TO N: READ W(I): NEXT I
2020 FOR I=1 TO N
2030   FOR J=1 TO N: READ F(I,J): NEXT J
2040 NEXT I
2050 FOR I=1 TO N                      ' Set [S] = [I]
2060   FOR J=1 TO N: S(I,J)=0: NEXT J
2070   S(I,I)=1
2080 NEXT I
2090 RETURN

3000 FOR I=1 TO N                      ' Update [D]
3010   FOR J=1 TO N
```

```
3020     Z=0
3030     FOR K=1 TO N: Z=Z+F(I,K)*W(K)*S(K,J): NEXT K
3040     D(I,J)=Z/GRAV
3050    NEXT J
3060 NEXT I
3070 FOR I=1 TO N: U(I)=1: NEXT I          ' Set {U} = {1}
3080 RETURN

4000 DIFF=1                               ' Stodola iteration
4010 PRINT : PRINT MODE
4020 WHILE DIFF>N*.000001
4030    FOR I=1 TO N
4040      Z=0
4050      FOR J=1 TO N: Z=Z+D(I,J)*U(J): NEXT J
4060      V(I)=Z
4070    NEXT I
4080    BIGV=0
4090    FOR I=1 TO N
4100      IF ABS(V(I))>ABS(BIGV) THEN BIGV=V(I)
4110    NEXT I
4120    DIFF=0
4130    FOR I=1 TO N
4140      Z=V(I)/BIGV
4150      DIFF=DIFF+ABS(Z-U(I))
4160      U(I)=Z
4170    NEXT I
4180 PRINT USING E5$; DIFF
4190 WEND
4200 RETURN

5000 UTWV=0                        ' Get mode properties
5010 VTWV=0
5020 FOR I=1 TO N
5030    WV=W(I)*V(I)
5040    UTWV=UTWV+U(I)*WV
5050    VTWV=VTWV+V(I)*WV
5060 NEXT I
5070 OMSQ=UTWV/VTWV
5080 FREQ(MODE)=SQR(OMSQ)
5090 PERIOD(MODE)=2*PI/FREQ(MODE)
5100 FOR I=1 TO N: PHI(I,MODE)=U(I): NEXT I
5110 WPHI=0: WPHISQ=0
5120 FOR I=1 TO N
5130    Z=W(I)*U(I)
```

```
5140    WPHI=WPHI+Z
5150    WPHISQ=WPHISQ+Z*U(I)
5160 NEXT I
5170 GAMMA(MODE)=WPHI/WPHISQ
5180 EQUIVWT(MODE)=GAMMA(MODE)*WPHI
5190 RETURN


6000 FOR I=1 TO N                        ' Update [S]
6010   FOR J=1 TO N
6020     S(I,J)=S(I,J)-U(I)*U(J)*W(J)/WPHISQ
6030   NEXT J
6040 NEXT I
6050 RETURN


7000 CLS: PRINT "[PHI]"                  ' Print results
7010 FOR I=1 TO N
7020   FOR J=1 TO MODES: PRINT USING F4$; FREQ(J);:
     NEXT J
7030 PRINT
7040 NEXT I
7050 PRINT: PRINT "Frequency, rad/sec"
7060 FOR J=1 TO MODES: PRINT USING F4$; PERIOD(J);:
     NEXT J
7070 PRINT: PRINT: PRINT "Period, sec "
7080 FOR J=1 TO MODES: PRINT USING F4$; PERIOD(J);:
     NEXT J
7090 PRINT: PRINT: PRINT "Modal participation factor"
7100 FOR J=1 TO MODES: PRINT USING F4$; GAMMA(J);:
     NEXT J
7110 PRINT: PRINT:
     PRINT "Base shear equivalent weight, kips"
7120 FOR J=1 TO MODES: PRINT USING F2$; EQUIVWT(J);:
     NEXT J
7130 END


8000 DATA 5, 5
8010 DATA 140, 120, 120, 120, 100
8020 DATA .0025, .0025, .0025, .0025, .0025
8030 DATA .0025, .0050, .0050, .0050, .0050
8040 DATA .0025, .0050, .0100, .0100, .0100
8050 DATA .0025, .0050, .0100, .0150, .0150
8060 DATA .0025, .0050, .0100, .0150, .0250
```

TABLE 6.3　Computed Modes for Structure of Fig. 6.1

Mode Number	1	2	3	4	5
$\{\phi\}$	$\begin{Bmatrix} 0.16806 \\ 0.32412 \\ 0.59657 \\ 0.79600 \\ 1.00000 \end{Bmatrix}$	$\begin{Bmatrix} -0.40590 \\ -0.64190 \\ -0.65330 \\ -0.19592 \\ 1.00000 \end{Bmatrix}$	$\begin{Bmatrix} -0.77985 \\ -0.86329 \\ 0.29142 \\ 1.00000 \\ -0.64456 \end{Bmatrix}$	$\begin{Bmatrix} -0.50579 \\ -0.14931 \\ 1.00000 \\ -0.77320 \\ 0.19975 \end{Bmatrix}$	$\begin{Bmatrix} -0.94886 \\ 1.00000 \\ -0.33752 \\ 0.09196 \\ -0.01190 \end{Bmatrix}$
ω (rad/sec)	8.8748	21.4880	31.3860	43.3656	58.0412
T (sec)	0.7080	0.2924	0.2002	0.1449	0.1082
Γ	1.4005	-0.5946	-0.3530	-0.1773	-0.1668
Base-shear equiv. weight (kips)	461.496	80.720	43.162	7.366	7.257
Cycles to converge	9	18	20	22	3

6.7 THE HOLZER METHOD FOR SHEARBUILDINGS

The defining attribute of a shearbuilding is that the shear in each story depends only on the displacement within that story. Hence, if the mass is taken to be lumped at the floor levels, the restoring force on any mass depends only upon the displacements of that mass and its immediate neighbors. Displacements of more remote masses have no effect. Such a system is said to be close-coupled, and the Holzer method is a simple and efficient way of finding its modes.

Dynamically, the shearbuilding is equivalent to a series of spring–mass systems connected end to end, or to a rotating shaft with flywheels—the system for which the Holzer process was devised originally. The crucial mathematical properties are that the inertia matrix $[M]$ is diagonal and the stiffness matrix $[K]$ is tridiagonal.

Assume that a shearbuilding is oscillating sinusoidally at a frequency ω, the top-floor absolute displacement being $z_n \sin \omega t$. The inertia force at the top floor is then $m_n \omega^2 z_n \sin \omega t$, which must be equal to the shear in the top story. Knowing the stiffness of the top story, we can calculate the displacement within the story, which in turn gives us the displacement of the $(n-1)$st floor. We then calculate its inertia force; dynamic equilibrium gives us the shear in the next lower story; from its stiffness, we calculate the story displacement to get the displacement of the next floor down; etc. We continue this down to the base, finally obtaining the amplitude of the base displacement. It is that base oscillation that drives the system. If the base displacement is zero, the frequency ω is a natural frequency and the displaced shape of the structure is the corresponding mode shape.

The structure of Fig. 6.1 will illustrate the process. Take the top-floor displacement amplitude to be $z_5 = 1$ in and take ω^2 to be 1200 sec^{-2}, which is $\omega = 34.641$ rad/sec. At peak displacement, the inertia force is $m_5\omega^2 z_5 = 310.81$ kips. The shear in the top story is thus 310.81 kips, and the displacement within the top story must be $\Delta_5 = 310.81$ kips/(100 kips/in) = 3.1081 in. Hence, the fourth-floor displacement is $z_4 = z_5 - \Delta_5 = 1.0000 - 3.1081 = -2.1081$ in. We work our way down the structure floor by floor, finally obtaining the base displacement $z_0 = -1.1570$ in. Table 6.4 shows the calculations. If the ω^2 we assumed, 1200 sec^{-2}, corresponded exactly to any one of the natural frequencies, then the base displacement would be zero. From the pattern of floor displacements, we can infer that 1200 sec^{-2} is somewhat greater than the third mode ω^2.

TABLE 6.4 Holzer Method Calculations

Floor	Floor weight (kips)	Floor displ. (in)	Inertia force (kips)	Story shear (kips)	Story stiffness (kips/in)	Story defl. (in)
			$\omega^2 = 1200$ sec^{-2} $\omega = 34.641$ rad/sec			
5	100	1.0000	310.81			
				310.81	100	3.1081
4	120	−2.1081	−786.26			
				−475.45	200	−2.3773
3	120	0.2692	100.39			
				−375.06	200	−1.8753
2	120	2.1445	799.83			
				424.77	400	1.0619
1	140	1.0826	471.06			
				895.82	400	2.2396
0		−1.1570				

The base-displacement amplitude is a function of ω^2, and the zeros of that function give the natural frequencies of the system. To find the zeros, we can set up an iteration based on Newton's method. Figure 5.9 shows the iteration for Newton's method, which employs the values of both the function and its first derivative at some approximate root to get an improved approximation. While we have no convenient way of evaluating the derivative, we can use a finite difference approximation:

$$f'(\omega^2) \simeq \{f(\omega^2 + \delta\omega^2) - f(\omega^2)\}/\delta\omega^2$$

In this case, repeating the process of Table 6.4 for $\omega^2 = 1210$ sec^{-2} yields $z_0 = -1.2144$ in. A pseudo-Newton iteration using a finite difference approximation of the derivative gives an improved approximation of the root as

$$\omega^2 = 1200 - f(1200) * (1210 - 1200)/\{f(1210) - f(1200)\}$$
$$= 1200 - (-1.1570) * 10/(-1.2144 + 1.1570)$$
$$= 998.432$$

Repeating the process leads rapidly to the mode, for which $\omega^2 = 985.082$ sec^{-2}. The final calculation is that shown in Table 6.5. The floor displacements divided by the extreme floor displacement give the mode shape, which turns out to be the third mode. The mode shape and frequency are, of course, identical with those obtained by the Stodola process, Table 6.3.

TABLE 6.5   Final Holzer Cycle for Third Mode

$$\omega^2 = 985.082 \text{ sec}^{-2} \qquad \omega = 31.386 \text{ rad/sec}$$

Floor	Floor weight (kips)	Floor displ. (in)	Inertia force (kips)	Story shear (kips)	Story stiffness (kips/in)	Story defl. (in)
5	100	1.0000	255.14			
				255.14	100	2.5514
4	120	−1.5514	−475.01			
				−219.87	200	−1.0993
3	120	−0.4521	−138.42			
				−358.29	200	−1.7915
2	120	1.3393	410.07			
				51.78	400	0.1295
1	140	1.2099	432.18			
				483.96	400	1.2099
0		0.0000				

The Holzer process enjoys a distinct advantage over the Stodola process in that any mode is calculated independently of the calculations for the other modes. With Stodola, the modes must be calculated in sequence, and any error in one mode propagates to all subsequent modal calculations. With Holzer, any mode can be sought independently of the others, in any order, and there is no inherited error. As with other Newton's method iterations, however, the process will not necessarily converge to the frequency sought. Whereas Stodola can be programmed for automatic calculation of any number of modes, Holzer, because of the possibility of nonconvergence or convergence to the wrong mode, is better suited to an interactive program in which the user provides the initial approximate frequency and then watches the results to see whether they converge to the desired mode.

Program 6.2 is an interactive Holzer program for shearbuildings. System properties are read in as data and the program then calls for the user to provide an approximate frequency. The computer then executes the calculations corresponding to Table 6.4 for two closely spaced values of ω^2, performs a pseudo-Newton extrapolation to an improved approximation, and repeats the process until either the base displacement or the difference between two successive approximate roots is negligibly small. It then divides the vector by its largest element, records the results, and asks the user to provide the approximate frequency for the next mode.

Program 6.2: Interactive Holzer Program for Shearbuildings

```
10    I3$="  ### "                          ' Housekeeping instr
20    E4$=" ##.####^^^^ "
30    E5$="  ##.#####^^^^ "
40    F2$="  ####.## "
50    F5$="  ####.##### "
60    GRAV=386.088
70    CLS: INPUT "Output device"; O$
      ' SCRN: for screen, LPT1: for printer,
      ' or D:FILENAME for file
80    OPEN O$ FOR OUTPUT AS #1

1000  READ N                               ' Get data
1010  DIM W(N), S(N), PHI(N)
1020  FOR I=1 TO N: READ W(I): NEXT I
1030  FOR I=1 TO N: READ S(I): NEXT I
1040  PRINT #1,                            ' Echo input data
1050  FOR I=1 TO N
1060     PRINT #1, USING I3$; I;
1070     PRINT #1, USING F2$; W(I); S(I)
1080  NEXT I
1090  PRINT #1,

2000  INPUT "Trial frequency, rad/sec"; OMEGA
         ' Trial frequency  (<=0 ends program)
2010  IF (OMEGA<=0) THEN CLOSE #1: END
2020  OMSQ=OMEGA^2
2030  TEST=OMSQ/65536!
2040  DF=TEST+1
2050  PRINT

3000  WHILE ABS(DF)>TEST          ' Pseudo-Newton iteration
3010     F1=OMSQ
3020     EPS=OMSQ/512
3030     GOSUB 6000
3040     Y1=X
```

```
3050    IF Y1=0 THEN DF=0: GOTO 3090
3060    OMSQ=F1+EPS
3070    GOSUB 6000
3080    DF=Y1*EPS/(X-Y1)
3090    OMSQ=F1-DF
3100    PRINT USING E4$; Y1; F1; DF
3110 WEND

4000 OMEGA=SQR(OMSQ)                    ' Get final mode shape
4010 GOSUB 6000
4020 PRINT USING E4$; X; OMSQ
4030 BIGPHI=0
4040 FOR I=1 TO N
4050    IF ABS(PHI(I))>ABS(BIGPHI) THEN BIGPHI=PHI(I)
4060 NEXT I
4070 FOR I=1 TO N: PHI(I)=PHI(I)/BIGPHI: NEXT I

5000 PRINT #1,: PRINT #1,;              ' Print results
5010 PRINT #1,"Omega = ";OMEGA
5020 PRINT #1,
5030 FOR I=1 TO N
5040    PRINT #1, USING F5$; PHI(I)
5050 NEXT I
5060 PRINT #1,
5070 GOTO 2000

6000 V=0                                ' Holzer calculations
6010 X=1
6020 FOR I=N TO 1 STEP -1
6030    PHI(I)=X
6040    V=V+(W(I)*OMSQ/GRAV)*X
6050    X=X-V/S(I)
6060 NEXT I
6070 RETURN

8000 DATA 5
8010 DATA 140,120,120,120,100
8020 DATA 400,400,200,200,100
```

Program 6.2 can illustrate some of the convergence peculiarities of pseudo-Newton iteration. For the given data, if we begin with a trial frequency of 37 rad/sec, which is between the third- and fourth-mode frequencies, the iteration converges to the third mode. With an initial trial of 38 rad/sec, it converges to the second mode; 39 leads to the first mode; 39.5 to the fifth; and 40 to the fourth. An initial 39.15783 rad/sec produces an overflow.

6.8 EXTENDED HOLZER METHOD

The Holzer method is readily extended to the general close-coupled system, for which $[M]$ or $[K]$ or both may be tridiagonal. For a linkage consisting of a series of rigid bars supported on springs, for example, both $[M]$ and $[K]$ could be tridiagonal.

  The process is fundamentally the same as before, although it is interpreted in concepts of mathematics instead of mechanics. We choose a trial value of ω^2 and determine the associated matrix $[A] = [[K] - \omega^2[M]]$. Because $[K]$ and $[M]$ are tridiagonal, $[A]$ is also tridiagonal. If the trial ω^2 corresponded to a true frequency and $\{\phi\}$ were proportional to the associated mode, we would have $[A]\{\phi\} = \{0\}$, where $\{\phi\} \neq \{0\}$. We assign a unit value to ϕ_1, get ϕ_2 from the first row of $[A]\{\phi\} = \{0\}$, use these two values to get ϕ_3 from the second row, and work down the matrix in that manner. The next-to-last row of $[A]\{\phi\} = \{0\}$ gives ϕ_n, and if the ω^2 were true, the last row would check. The first $n - 1$ equations are satisfied, and the discrepancy in the nth equation is a measure of the error. It plays the same role as the base amplitude in the Holzer method for shearbuildings — it would be zero if the ω^2 corresponded to a true natural frequency.

  In the Holzer method for shearbuildings, we worked from top to bottom. We could here as well, assigning unit value to ϕ_n and finding the remaining ϕ's in reverse order. The end results would be the same.

  We can modify Program 6.2 to execute the extended Holzer process by replacing the input instructions 1000–1090, the Holzer subroutine 6000–6070, and the data statements 8000–8020. Program 6.3 gives replacements. We store the diagonal elements of $[K]$ as $k_1, k_2, \ldots, k_n$, and the off-diagonal elements as $k1_1, \ldots, k1_{n-1}$. $[M]$ and $[A]$ are stored similarly.

Program 6.3: Interactive Holzer Program for General Close-Coupled System; Tridiagonal [*M*] and [*K*]

```
[Merge with Program 6.2 for the complete interactive
program.]
1000 READ N                          ' Get data
1010 DIM M(N), M1(N-1), K(N), K1(N-1),
     PHI(N), A(N), A1(N-1)
1020 FOR I=1 TO N: READ M(I): NEXT I
1030 FOR I=1 TO N-1: READ M1(I): NEXT I
1040 FOR I=1 TO N: READ K(I): NEXT I
1050 FOR I=1 TO N-1: READ K1(I): NEXT I
1060 A1(0)=0
1070 K1(0)=0
```

```
1080 M1(0)=0
1090 PHI(0)=0
1100 PRINT #1,                        ' Echo input data
1110 FOR I=1 TO N-1
1120    PRINT #1, USING I3$; I;
1130    PRINT #1, USING F5$; M(I); M1(I);
1140    PRINT #1, USING F2$; K(I); K1(I)
1150 NEXT I
1160 PRINT #1, USING I3$; N;
1170 PRINT #1, USING F5$; M(N);
1180 PRINT #1, SPC(12);
1190 PRINT #1, USING F2$; K(N)
1200 PRINT #1,

6000 FOR I=1 TO N-1                    ' Holzer calculations
6010    A(I)=K(I)-OMSQ*M(I)
6020    A1(I)=K1(I)-OMSQ*M1(I)
6030 NEXT I
6040 A(N)=K(N)-OMSQ*M(N)
6050 PHI(1)=1
6060 FOR I=1 TO N-1
6070    PHI(I+1)=-(A1(I-1)*PHI(I-1)+A(I)*PHI(I))/A1(I)
6080 NEXT I
6090 X=A1(N-1)*PHI(N-1)+A(N)*PHI(N)
6100 RETURN

8000 DATA 5
8010 DATA .36261,.31081,.31081,.31081,.25901
8020 DATA 0,0,0,0
8030 DATA 800,600,400,300,100
8040 DATA -400,-200,-200,-100
```

6.9 MODES FROM THE FREQUENCY EQUATION

In Chapter 5, we derived the frequency equation:

$$\|[K] - \omega^2[M]\| = 0 \qquad (5.6.12)$$

which we expanded into a polynomial in ω^2, the frequency polynomial. Its roots were the natural frequencies of the system, and associated with each root ω^2 was a mode $\{\phi\}$. The approach was valid, but we rejected it because of the difficulty of expanding the determinant into a polynomial and then finding its roots, except for systems having only a few degrees of freedom. Let us now revive the concept using an alternative approach. Let

$[A] = [[K] - \omega^2[M]]$. The determinant of $[A]$ is then a function of ω^2, and we will seek its roots without expanding it into a polynomial. For this we need an efficient procedure for finding the determinant of $[A]$ and for solving the equation $[A]\{\phi\} = 0$.

6.10 DETERMINANT EVALUATION

The matrix $[A]$, being symmetrical, can be resolved into factors $[T]$ and $[D]$ such that

$$[T]^T[D][T] = [A] \qquad (6.10.1)$$

where $[D]$ is a diagonal matrix, and $[T]$ is a unit upper triangular matrix, having all its diagonal elements equal to 1 and all elements below the diagonal equal to zero. The determinant of $[A]$ is then equal to the determinant of $[D]$, which is simply the product of its diagonal elements.

For a system of n degrees of freedom, there are $n(n + 1)/2$ independent elements of $[A]$. There are n elements of $[D]$ and $n(n - 1)/2$ off-diagonal elements of $[T]$. Thus, the number of elements of $[D]$ and $[T]$ to be determined is equal to the number of independent elements of $[A]$, and the factors are unique. Indeed, we can devise a simple algorithm for finding them. Equation (6.10.1) is equivalent to

$$\sum_{k=1}^{i} T_{ki} D_k T_{kj} = A_{ij} \qquad i = 1, 2, \ldots, n; j = i, i + 1, \ldots, n \qquad (6.10.2)$$

Because $[T]$ is upper triangular, the summation covers only the range $k \leq i$, and we need consider only the range $j \geq i$ for the same reason. The diagonal elements of $[T]$ are all equal to 1. Thus, for $i = 1$:

$$\begin{aligned} T_{11} &= 1 \\ D_1 &= A_{11} \\ T_{1j} &= A_{1j}/D_1 \qquad j = 2, 3, \ldots, n \end{aligned} \qquad (6.10.3)$$

For $i = 2$:

$$\begin{aligned} T_{22} &= 1 \\ D_2 &= A_{22} - T_{12}D_1T_{12} \\ T_{2j} &= (A_{2j} - T_{12}D_1T_{1j})/D_2 \qquad j = 3, 4, \ldots, n \end{aligned} \qquad (6.10.4)$$

In general, we can compute for each value of i in $i = 1, 2, \ldots, n$:

$$T_{ii} = 1$$

$$D_i = A_{ii} - \sum_{k=1}^{i-1} T_{ki} D_k T_{ki}$$

$$T_{ij} = \left(A_{ij} - \sum_{k=1}^{i-1} T_{ki} D_k T_{kj}\right)\Big/D_i \quad j = i + 1, i + 2, \ldots, n \tag{6.10.5}$$

The order of computation is important. We must compute all the elements for each value of i before proceeding to the next, in order to have each element of $[T]$ and $[D]$ before it appears in a summation.

Having the matrix factors $[T]$ and $[D]$, we can compute the determinant of $[A]$ as the product of the diagonal elements of $[D]$.

6.11 SOLUTION OF LINEAR EQUATIONS

We need also a convenient process for solving linear equations, and the matrix factors lead to a simple algorithm. Consider the equation

$$[A]\{x\} = \{b\} \tag{6.11.1}$$

where $[A]$ is symmetrical. Resolving $[A]$ into factors $[T]$ and $[D]$ as in Sec. 6.10, we have

$$[T]^T[D][T]\{x\} = \{b\} \tag{6.11.2}$$

We employ an intermediate vector $\{v\}$ such that

$$\{v\} = [T]\{x\} \tag{6.11.3}$$

Then

$$[T]^T[D]\{v\} = \{b\} \tag{6.11.4}$$

In subscript notation, this is

$$\sum_{k=1}^{i} T_{ki} D_k v_k = b_i \tag{6.11.5}$$

Because $[T]$ is triangular, the summation covers only the range $k = 1$ to $k = i$, and we can determine $\{v\}$ from Eq. (6.11.5) in the following manner.

For $i = 1$:

$$T_{11} D_1 v_1 = b_1, \quad \text{or,} \quad \text{since } T_{11} = 1, \quad v_1 = b_1/D_1 \tag{6.11.6}$$

For $i = 2$:

$$T_{12}D_1v_1 + T_{22}D_2v_2 = b_2, \quad \text{or} \quad v_2 = (b_2 - T_{12}D_1v_1)/D_2 \qquad (6.11.7)$$

In general,

$$v_i = \left(b_i - \sum_{k=1}^{i-1} T_{ki}D_k v_k\right)\Big/D_i \qquad i = 1, 2, \ldots, n \qquad (6.11.8)$$

We must compute the v_i's in increasing order so that each element is known when it is needed in the summation.

Now, having $\{v\}$, we find $\{x\}$ from Eq. (6.11.3), which in subscript notation is

$$\sum_{j=i}^{n} T_{ij}x_j = v_j \qquad (6.11.9)$$

The summation runs only from $j = i$ to $j = n$ because $[T]$ is an upper triangle, and we must go in decreasing order of i.

For $i = n$:

$$T_{nn}x_n = v_n, \quad \text{or,} \quad \text{since } T_{nn} = 1, \, x_n = v_n \qquad (6.11.10)$$

For $i = n - 1$:

$$T_{n-1, n-1}x_{n-1} + T_{n-1, n}x_n = v_{n-1}$$

or

$$x_{n-1} = v_{n-1} - T_{n-1, n}x_n \qquad (6.11.11)$$

In general,

$$x_i = v_i - \sum_{j=i+1}^{n} T_{ij}x_j \qquad i = n, n - 1, \ldots, 1 \qquad (6.11.12)$$

6.12 DETERMINANT AND LINEAR-EQUATIONS SUBROUTINE

The subroutine of Program 6.4 uses the algorithms of Secs. 6.10 and 6.11 to evaluate the determinant of $[A]$ and solve the linear equations $[A]\{x\} = \{b\}$. It first finds the factors $[D]$ and $[T]$ and the intermediate vector $\{v\}$, then calculates the determinant, and finally gets the solution vector $\{x\}$.

Program 6.4: Subroutine for Finding the Determinant of [A] and the Solution Vector {x}, Given the Equation [A] {x} = {b}, with [A] a Symmetrical Matrix

```
100 FOR I=1 TO N
101   T(I,I)=1
102   FOR J=I TO N
103     SUM=A(I,J)
104     FOR P=1 TO I-1
105       SUM=SUM-T(P,I)*D(P)*T(P,J)
106     NEXT P
107     IF J=I THEN D(I)=SUM ELSE T(I,J)=SUM/D(I)
108   NEXT J
109   SUM=B(I)
110   FOR P=1 TO I-1
111     SUM=SUM-T(P,I)*D(P)*V(P)
112   NEXT P
113   V(I)=SUM/D(I)
114 NEXT I

200 DET=1
201 FOR I=1 TO N: DET=DET*D(I): NEXT I

300 FOR I=N TO 1 STEP -1
301   SUM=V(I)
302   FOR J=I+1 TO N
303     SUM=SUM-T(I,J)*X(J)
304   NEXT J
305   X(I)=SUM
306 NEXT I
307 RETURN
```

Statements 100–114 determine $[D]$, $[T]$, and $\{v\}$. The algorithms for D_i and T_{ij} are identical except for the divisor D_i that appears in one but not the other. Statements 102–108 compute both, comparing i and j in statement 107 to determine whether the result is D_i ($i = j$) or T_{ij} ($i \neq j$).

Statements 200–201 get the determinant, and statements 300–306 compute $\{x\}$.

Program 6.4 can be improved both in efficiency and in memory requirements by taking advantage of the following observations:

A_{ii} is never used after D_i is calculated.

A_{ij} ($j > i$) is never used after T_{ij} is calculated.

$T_{ii} = 1$ for all i; hence, there is no need to record it.

b_i is never used after v_i is calculated.

v_i is never used after x_i is calculated.

The b_i-to-v_i algorithm (statements 109–113) is identical with the A_{ij}-to-T_{ij} algorithm (statements 103–107).

To utilize these properties, we first augment the matrix [A] by appending the vector $\{b\}$ as the $(n + 1)$st column. Then we omit storing the diagonal elements T_{ii}, overlay the D_i's on the diagonal elements of $[A \mid b]$, and overlay the T_{ij}'s, v_i's, and x_i's on the off-diagonal elements of $[A \mid b]$. The subroutine of Program 6.5 executes the process.

Program 6.5: Compact Subroutine for Finding the Determinant of a Symmetrical Matrix [A] and the Solution Vector $\{x\}$ for the Equation $[A]\{x\} = \{b\}$

```
100 FOR I=1 TO N
101   FOR J=I TO N+1
102     SUM=A(I,J)
103     FOR P=1 TO I-1
104       SUM=SUM-A(P,I)*A(P,P)*A(P,J)
105     NEXT P
106     IF J=I THEN A(I,I)=SUM ELSE A(I,J)=SUM/A(I,I)
107   NEXT J
108 NEXT I

200 DET=1
201 FOR I=1 TO N: DET=DET*A(I,I): NEXT I

300 FOR I=N-1 TO 1 STEP -1
301   SUM=A(I,N+1)
302   FOR J=I+1 TO N
303     SUM=SUM-A(I,J)*A(J,N+1)
304   NEXT J
305   A(I,N+1)=SUM
306 NEXT I

400 RETURN
```

Upon completion of Program 6.5, the variable DET is the determinant of [A] and the $(n + 1)$st column of [A] is the solution vector $\{x\}$. The original matrix [A] and the vector $\{b\}$ are destroyed. The lower triangle of [A] is not used at all, and is left undisturbed.

6.13 INTERACTIVE DETERMINANT PROGRAM

Using these components, we can now assemble an interactive program to find the frequencies and modes for a dynamic system. The procedure will be much the same as the Holzer interactive Programs 6.2 and 6.3, except that this time we seek the zeros of a determinant rather than the zero base amplitude or zero error in an equation. In this case, the system need not be close-coupled. $[M]$ or $[K]$ or both may be full matrices.

We use a pseudo-Newton iteration process to find the frequencies. Starting with a trial frequency ω, we set up the dynamic matrix $[A] = [[K] - \omega^2[M]]$ and use the determinant portion of the subroutine of Program 6.5 to find the determinant of $[A]$. We then add a small increment to ω^2, set up a new dynamic matrix $[A]$, and find its determinant. Using these two determinant values, we project linearly to find the value of ω^2 corresponding to a zero determinant. We take this as an improved approximation of ω^2 and repeat the process, continuing until either the determinant is zero or the change in ω^2 from one cycle to the next is negligibly small. If the initial trial is reasonably close to a natural frequency, the process ought to converge fairly rapidly. Again there is the possibility of nonconvergence or convergence to a natural frequency other than the one sought. Hence, an interactive program is advantageous, for it lets the user watch the successive approximations to see that things are not going astray.

Once the frequency is obtained, we find the mode shape by solving the equations $[A]\{\phi\} = 0$. Of course, $[A]$ is singular and the solution vector $\{\phi\}$ is not unique. We therefore assign an arbitrary value to one element of $\{\phi\}$ and solve for the remaining elements. A convenient approach is to assign a value $\phi_n = -1$, and find the other ϕ's from the first $n - 1$ equations. Each equation is of the form

$$a_{11}\phi_1 + a_{12}\phi_2 + a_{13}\phi_3 + \ldots + a_{1n}\phi_n = 0 \qquad (6.12.1)$$

or, since $\phi_n = -1$,

$$a_{11}\phi_1 + a_{12}\phi_2 + \ldots + a_{1,n-1}\phi_{n-1} = a_{1n} \qquad (6.12.2)$$

Thus, the first $n - 1$ rows of $[A]$ are set up exactly in the form required for solution as a system of $n - 1$ equations by the subroutine of Program 6.5.

Program 6.6 reads the input data, accepts a trial frequency provided by the user and performs a pseudo-Newton iteration to get the frequency for a zero determinant. It prints successive values of the determinant, ω^2, and the change of ω^2 on the screen to let the user observe the behavior of the iteration. Once it obtains a root ω^2, it assigns a value of -1 to ϕ_n, solves for the remaining ϕ's, and normalizes them to get the mode shape. Finally, it prints the frequency and mode shape and then asks for a new trial frequency.

Program 6.6: Interactive Program for Obtaining Frequencies and Modes from the Frequency Equation

```
[Merge with Program 6.2 for the complete interactive
program.]
1000 SWITCH=0
1010 READ N                              ' Get input data
1020 DIM M(N,N),K(N,N),A(N,N),PHI(N)
1030 FOR I=1 TO N
1040   FOR J=I TO N: READ M(I,J): M(J,I)=M(I,J): NEXT J
1050 NEXT I
1060 FOR I=1 TO N
1070   FOR J=I TO N: READ K(I,J): K(J,I)=K(I,J): NEXT J
1080 NEXT I
1090 PRINT #1,: PRINT #1, "[M]"          ' Echo input data
1100 FOR I=1 TO N
1110   FOR J=1 TO N: PRINT #1, USING F5$; M(I,J);:
       NEXT J
1120 PRINT #1,
1130 NEXT I
1140 PRINT #1,: PRINT #1, "[K]"
1150 FOR I=1 TO N
1160   FOR J=1 TO N: PRINT #1, USING F5$; K(I,J);:
       NEXT J
1170 PRINT #1,
1180 NEXT I
1190 PRINT #1,

4000 SWITCH=1                            'Compute mode
4010 OMEGA=SQR(OMSQ)
4020 PRINT: PRINT "Frequency = "; OMEGA;
4030 GOSUB 6000
4040 GOSUB 7000
4050 DISCR=0
4060 FOR J=1 TO N
4070   DISCR=DISCR+A(N,J)*PHI(J)
4080 NEXT J
4090 PRINT "  Phi Error = "; DISCR: PRINT
4100 SWITCH=0

6000 FOR I=1 TO N                ' Get components of [A]
6010   FOR J=1 TO N
6020     A(I,J)=K(I,J)-OMSQ*M(I,J): A(J,I)=A(I,J)
6030   NEXT J
6040 NEXT I
```

```
6050 FOR I=1 TO N-SWITCH
6060   FOR J = I TO N
6070     SUM = A(I,J)
6080     FOR P=1 TO I-1
6090       SUM = SUM - A(P,I)*A(P,P)*A(P,J)
6100     NEXT P
6110     IF I=J THEN A(I,I)=SUM ELSE A(I,J) = SUM/A(I,I)
6120   NEXT J
6130 NEXT I
6140 IF SWITCH=1 THEN RETURN
6150 X=1                          ' Find determinant of [A]
6160 FOR I=1 TO N: X=X*A(I,I): NEXT I
6170 RETURN

7000 FOR I=N-2 TO 1 STEP -1            ' Final mode shape
7010   SUM=A(I,N)
7020   FOR J=I+1 TO N-1
7030     SUM=SUM-A(I,J)*A(J,N)
7040   NEXT J
7050   A(I,N)=SUM
7060 NEXT I
7070 BIGPHI=-1
7080 FOR I=1 TO N-1
7090   IF ABS(A(I,N))>ABS(BIGPHI) THEN BIGPHI=A(I,N)
7100 NEXT I
7110 PHI(N)=-1/BIGPHI
7120 FOR I=1 TO N-1
7130   PHI(I)=A(I,N)/BIGPHI
7140 NEXT I
7150 RETURN

8000 DATA 5                  ' Data for structure of Fig. 6.1
8010 DATA .36261,0,0,0,0
8020 DATA .31081,0,(,0
8030 DATA .31081,0,0
8040 DATA .31081,0
8050 DATA .25901
8060 DATA 800,-400,0,0,0
8070 DATA 600,-200,0,0
8080 DATA 400,-200,0
8090 DATA 300,-100
8100 DATA 100
```

The variable SWITCH (set by statement 1000, 4000, or 4100) governs which of two purposes subroutine 6000–6170 performs. If SWITCH = 0, the subroutine computes the determinant of [A] for the pseudo-Newton iteration. If SWITCH = 1, the subroutine solves the first $n - 1$ equations of

$[A]\{\phi\} = 0$, with a value of -1 assigned to ϕ_n, then divides $\{\phi\}$ by its largest element, and finally computes the error in the unused nth equation of $[A]\{\phi\} = 0$. The error serves as an indicator of how good the mode shape is, and, of course, a huge error would indicate that there is a mistake in the program somewhere.

Table 6.6 shows the program output for one mode, the third mode in this case. The output to the selected output device (screen, printer, or disk file) is shown in roman type, and the screen output is shown in italics. Only an examination of the mode shape reveals which mode has been found.

TABLE 6.6   Output of Program 6.6 for Example Structure

[M]				
0.36261	0.00000	0.00000	0.00000	0.00000
0.00000	0.31081	0.00000	0.00000	0.00000
0.00000	0.00000	0.31081	0.00000	0.00000
0.00000	0.00000	0.00000	0.31081	0.00000
0.00000	0.00000	0.00000	0.00000	0.25901

[K]				
800.00000	−400.00000	0.00000	0.00000	0.00000
−400.00000	600.00000	−200.00000	0.00000	0.00000
0.00000	−200.00000	400.00000	−200.00000	0.00000
0.00000	0.00000	−200.00000	300.00000	−100.00000
0.00000	0.00000	0.00000	−100.00000	100.00000

Trial frequency, rad/sec? 30

2.0905E+11	*9.0000E+02*	*−1.0283E+02*
−5.1976E+10	*1.0028E+03*	*1.7285E+01*
−1.3295E+09	*9.8555E+02*	*4.6368E−01*
−4.6815E+06	*9.8508E+02*	*1.6350E−03*

Frequency = 31.38603   Phi Error = 1.525879E−05

Omega = 31.38603

−0.77985
−0.86329
0.29141
1.00000
−0.64455

Trial frequency, rad/sec?

6.14 STRUCTURAL APPROXIMATIONS AND CHOICE OF METHOD

We now have several methods of computing the modes of a dynamic system, each with its merits and drawbacks. The choice of method depends partly upon the approximations employed in modeling the structure mathematically.

The Stodola process is applicable to any system, but it requires inverting the stiffness matrix $[K]$ and it must compute the modes in sequence, starting with mode 1. This is not a real drawback, for the first few modes are ordinarily the important ones. The process can be modified to reverse the sequence, starting with the highest mode and working down, but that would rarely be useful.

The Holzer process is faster than Stodola and can find any mode independently of the others. Thus, error does not propagate from one mode calculation to the next. However, the process is limited to close-coupled systems in which the stiffness matrix $[K]$ and mass matrix $[M]$ are either diagonal or tridiagonal, and unlike Stodola, it cannot readily be set up for automatic calculation, but must be used interactively.

The frequency-equation determinant process escapes the tridiagonal limitation of Holzer and, like Stodola, can be used for any system. It shares the drawback of Holzer in being limited to interactive use, and it is much slower than Holzer. Like Holzer, it computes each mode independently of the rest, and thus avoids error propagated from one mode calculation to the next.

An approximation sometimes used in modeling a structure is the shearbuilding approximation, in which the beams of a framework are taken to be infinitely rigid compared with the columns. This is attractive because of its simplicity — it makes it very easy to calculate the stiffness matrix $[K]$, and $[M]$ is ordinarily taken to be diagonal. Thus, any of the mode computation processes could be used. The shearbuilding approximation could, however, produce misleading results. If the beams were in fact much stiffer than the columns, it might be acceptable, but for tall buildings, the columns might be as stiff as the beams or even stiffer. In such a case, neglecting beam flexibility would lead to computed frequencies far greater than would prevail if beam flexibility were taken into account, leading to smaller computed displacements and most likely to greater computed stresses for equal dynamic input forces.

The shearbuilding approximation introduces no discontinuities and no stress or strain incompatibilities. An n-story plane frame will have n valid modes, whether flexibility of the beams is considered or ignored, as long as discontinuities or incompatibilities are avoided. Other approximations may be troublesome. One tempting approximation might be simply to consider only the tridiagonal elements of $[K]$ and ignore the rest, thus to obtain a system for which the Holzer process could be used instead of Stodola or the frequency-equation process. Alas, the tridiagonal elements of $[K]$ do not make a valid stiffness matrix, and the process might or might not work. Mathematically, the dynamic flexibility matrix $[K]^{-1}[M]$ would still have n eigenvalues and eigenvectors, but some of the eigenvalues, which are the values of ω^2, might even be negative. Other less obviously invalid approximations might lead to similar difficulties.

The portal method, for example, is a classical method of calculating the approximate bending moments in frame members due to lateral forces. It assumes that there is a point of contraflexure at midheight of each column and at midspan of each beam, and that the shear in each story is shared by the columns, one portion to each interior column and one-half portion to each exterior column. These assumptions convert a cumbersome statically indeterminate problem to a statically determinate one. That is fine, as far as it goes. Difficulties arise if one tries to extend the process to calculating lateral displacements, which must be done to evaluate the stiffness matrix $[K]$. While the portal method stresses satisfy equilibrium, the strains violate compatibility. Different lateral displacements might be obtained, depending on how the compatibility violations are accommodated. One could, for example, compute the lateral displacements from the strains in the exterior columns, or one could compute them from the strains in the interior columns. The results would differ. Either would yield an invalid $[K]$.

True matrices $[K]$ and $[M]$ for an approximate structural system will produce valid modes and frequencies. Approximate matrices $[K]$ and $[M]$ for a true structural system might not.

6.15 EARTHQUAKE RESPONSE BY NORMAL MODES

Normal modes are useful in calculating the response of a multidegree-of-freedom system to earthquake. Consider the idealized five-story building of Fig. 6.1. Table 6.3 gives the modes. We now use response spectrum techniques to evaluate its response to an earthquake component, specifically, to the N11°W component of the Eureka, California, earthquake of December 21, 1954. Figure 4.3 shows the response spectra for this earthquake component. Table 6.1 gives the essential mode data. One additional set of building dimensions we need is the floor heights; we will take the story height to be a uniform 11 ft.

Taking damping to be 5 percent of critical damping for all modes, we get SD for each mode from Fig. 4.3 or the data from which it was plotted. The response is then computed from these relations:

Maximum modal displacement: $q_p = |\Gamma_p \, SD_p|$ (6.15.1)

Extreme floor displacement: $u_{ip} = \phi_{ip} q_p$ (6.15.2)

Extreme story drift: $\Delta_{ip} = u_{ip} - u_{i-1,p}$ (6.15.3)

Extreme inertia force: $F_{ip} = m_i \omega_p^2 u_{ip}$ (6.15.4)

Extreme story shear: $V_{ip} = \sum_{k=i}^{n} F_{kp}$ (6.15.5)

Base shear: Base $V_p = \sum_{i=1}^{n} F_{ip}$ (6.15.6)

or   Base $V_p = \omega_p^2 SD_p m_p^v$ (6.15.7)

Base overturning moment: $OTM_p = \sum_{i=1}^{n} F_{ip} h_i$ (6.15.8)

Table 6.7 shows the response for each of the five modes and the modal responses combined, where applicable, using both the sum of the absolute values and the square root of the sum of the squares.

TABLE 6.7   Seismic Response of Five-story Example Structure

Mode Number	1	2	3	4	5	SumAbs	RSS
T (sec)	0.708	0.292	0.200	0.145	0.108		
SD (in)	1.163	0.421	0.138	0.056	0.029		
Γ	1.401	−0.594	−0.353	−0.177	−0.167		
q (in)	1.63	0.25	0.05	0.01	0.005		
Base shear (kips)	109.5	40.7	15.2	2.0	1.8	169.2	117.8
Base OTM (kip-ft)	4378	132	74	10	11	4605	4381
Max. displ. (in) (flr)	1.63(5)	0.25(5)	0.05(4)	0.01(3)	0.005(2)	1.91(5)	1.65(5)
Max. drift (in) (sty)	0.44(3)	0.30(5)	0.08(5)	0.02(4)	0.01(2)	0.72(5)	0.45(5)

PROBLEMS

6.1 [Feasible for hand calculation.] Find the modes of the structure of Problem 5.6 by the Stodola process.

6.2 [Feasible for hand calculation.] Find the modes of the structure of Problem 5.6 by the Holzer method.

6.3 [Feasible for hand calculation.] Find the second mode of the structure of Problem 5.7 by the frequency-equation determinant evaluation method.

6.4 (a) [Tedious for hand calculation.] Find the modes of the spring–mass
system shown in the figure by the Stodola process.
(b) [Feasible for hand calculation.] Find the modes by the Holzer
method.

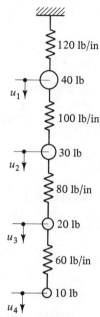

6.5 (a) [Tedious for hand calculation.] Find the modes of the spring–mass
system shown in the figure by the Stodola process.
[Note: With an initial trial of $\{U\} = \{1\}$, the Stodola process
may fail to converge to the even-numbered modes, which are antisymmetrical. To avoid this difficulty, set the initial $u_i = 1$ for the
left half and $u_i = -1$ for the right half when the mode number is
even.]
(b) [Tedious for hand calculation.] Find the modes by the extended
Holzer method.

	50 lb		100 lb		100 lb		50 lb	
100 lb/in		100 lb/in		100 lb/in		100 lb/in		100 lb/in
	u_1		u_2		u_3		u_4	

6.6 [Impractical for hand calculation.] A uniform cantilever beam has four masses attached, equally spaced, as shown in the figure. The beam properties are

$$E = 10,000,000 \text{ psi}$$
$$I = 2.00 \text{ in}^4$$
$$\text{Length} = 96 \text{ in}$$
$$\text{Weight} = \text{negligible}$$

Find the four modes and frequencies.

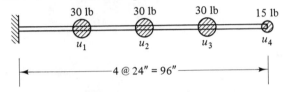

6.7 [Impractical for hand calculation.] A simply supported uniform beam has five 25-kip weights attached, equally spaced, as shown in the figure. The beam properties are

$$E = 29,500 \text{ ksi}$$
$$I = 1500 \text{ in}^4$$
$$\text{Length} = 30 \text{ ft}$$
$$\text{Weight} = \text{negligible}$$

Find the five modes and frequencies.

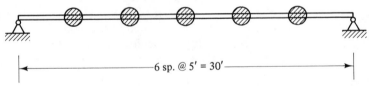

CHAPTER SEVEN

Continuous Systems

7.1 THE EQUATION OF UNDAMPED MOTION

Consider the undamped transverse vibration of a straight beam supported in such a way that its supports do not contribute to the strain energy of the system. This requires that the supports either prevent translation completely (displacement = 0) or offer no resistance to translation (force = 0), and that they either prevent rotation or offer no resistance to rotation. The customary supports—simple, fixed, guided, and free—all meet these requirements. We neglect shear deformation and rotatory inertia, and for ease in manipulating the equations, we take the supports to be at the ends of the beam.

Let

l = length of beam

x, y = longitudinal and transverse coordinates, respectively

v = transverse displacement

$EI(x)$ = flexural stiffness

$m(x)$ = mass per unit length

$f(x, t)$ = dynamic driving force per unit length

$\mathbf{V}(x, t)$ = transverse shear force

$\mathbf{M}(x, t)$ = bending moment

Figure 7.1 shows the beam and the forces on a differential element.

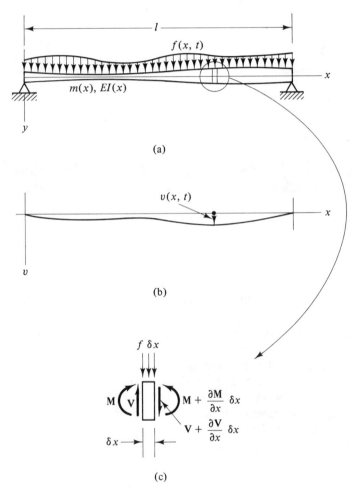

(a)

(b)

(c)

Figure 7.1 Continuous system. (a) Beam and driving force. (b) Displacement. (c) Forces on element.

The equation of motion for the element is

$$f \, \delta x + (\partial V / \partial x) \, \delta x = m(\partial^2 v / \partial t^2) \, \delta x \qquad (7.1.1)$$

With shear deformation neglected, the moment–curvature relation is

$$\mathbf{M} = -EI(\partial^2 v / \partial x^2) \qquad (7.1.2)$$

and, in the absence of rotatory inertia, rotational equilibrium of the element requires that

$$\mathbf{V} = \partial \mathbf{M} / \partial x \qquad (7.1.3)$$

Putting Eqs. (7.1.2) and (7.1.3) into Eq. (7.1.1), we get

$$m(\partial^2 v / \partial t^2) + \partial^2 (EI \partial^2 v / \partial x^2) / \partial x^2 = f(x, t) \qquad (7.1.4)$$

which is the partial differential equation of motion for the system.

For the case of free vibration, Eq. (7.1.4) becomes

$$m(\partial^2 v / \partial t^2) + \partial^2 (EI \partial^2 v / \partial x^2) / \partial x^2 = 0 \qquad (7.1.5)$$

We separate variables by letting

$$v(x, t) = \psi(x) q(t) \qquad (7.1.6)$$

Then

$$\partial^2 v / \partial t^2 = \psi \ddot{q}$$
$$\partial^2 v / \partial x^2 = \psi'' q \qquad (7.1.7)$$
$$\partial^2 (EI \partial^2 v / \partial x^2) / \partial x^2 = q(d^2 [EI\psi''] / dx^2)$$

where dots denote time derivatives and primes denote x derivatives. Now we substitute Eqs. (7.1.7) into Eq. (7.1.5) to get

$$m\psi \ddot{q} + q(d^2 [EI\psi''] / dx^2) = 0 \qquad (7.1.8)$$

which, when divided by $m\psi q$, becomes

$$\frac{\ddot{q}}{q} + \frac{d^2 [EI\psi''] / dx^2}{m\psi} = 0 \qquad (7.1.9)$$

The first expression in Eq. (7.1.9) is independent of x and the second is independent of t. To be valid for all values of x and t, the two expressions must therefore be constant, which leads us to

$$-\frac{\ddot{q}}{q} = \frac{d^2 [EI\psi''] / dx^2}{m\psi} = \text{constant} = \omega^2 \qquad (7.1.10)$$

Thus, the partial differential equation of undamped free vibration, Eq. (7.1.5), becomes two ordinary differential equations, the familiar time equation

$$\ddot{q} + \omega^2 q = 0 \tag{7.1.11}$$

and a deformation equation

$$d^2[EI\psi'']/dx^2 - \omega^2 m\psi = 0 \tag{7.1.12}$$

For any given stiffness and mass functions $EI(x)$ and $m(x)$, respectively, there is an infinite set of frequencies ω^2 and associated mode shapes $\psi(x)$ that satisfy the deformation equation and the support conditions for the beam.

7.2 EXAMPLE: UNIFORM SIMPLY SUPPORTED BEAM

For a uniform beam, EI is constant and Eq. (7.1.12) becomes

$$EI\psi^{iv} - \omega^2 m\psi = 0 \tag{7.2.1}$$

or

$$\psi^{iv} - \omega^2(m/EI)\psi = 0 \tag{7.2.2}$$

Let

$$\lambda^4 = \omega^2 m/EI \tag{7.2.3}$$

Then Eq. (7.2.2) becomes

$$\psi^{iv} - \lambda^4\psi = 0 \tag{7.2.4}$$

The general solution of Eq. (7.2.4) can be expressed in terms of circular and hyperbolic sines and cosines; thus,

$$\psi = C_1 \cos \lambda x + C_2 \sin \lambda x + C_3 \cosh \lambda x + C_4 \sinh \lambda x \tag{7.2.5}$$

For a beam simply supported at its two ends, the displacement and bending moment at the ends are zero. Thus, at end 0,

$$v(0) = 0 \rightarrow \quad \psi(0) = 0 \rightarrow \quad C_1 + C_3 = 0 \tag{7.2.6}$$

$$\mathbf{M}(0) = 0 \rightarrow EI\psi''(0) = 0 \rightarrow \lambda^2(-C_1 + C_3) = 0 \tag{7.2.7}$$

from which

$$C_1 = C_3 = 0 \tag{7.2.8}$$

Then, at end l,

$$v(l) = 0 \rightarrow \quad \psi(l) = 0 \rightarrow \quad C_2 \sin \lambda l + C_4 \sinh \lambda l = 0 \tag{7.2.9}$$

$$\mathbf{M}(l) = 0 \rightarrow EI\psi''(l) = 0 \rightarrow \lambda^2(-C_2 \sin \lambda l + C_4 \sinh \lambda l) = 0 \tag{7.2.10}$$

from which

$$C_2 \sin \lambda l = C_4 \sinh \lambda l = 0 \qquad (7.2.11)$$

Sinh λl cannot be zero, so C_4 must be zero. For a nontrivial solution, at least one of the C_i's must be nonzero, and C_2 is the only one left. Therefore, $\sin \lambda l$ must be zero, from which

$$\lambda l = n\pi \qquad n = 1, 2, \ldots \qquad (7.2.12)$$

Equation (7.2.3) then gives us

$$\omega_n = n^2 \pi^2 (EI/ml^4)^{1/2} \qquad (7.2.13)$$

and from Eq. (7.2.5), we get

$$\psi_n = C_2 \sin (n\pi x/l) \qquad (7.2.14)$$

The value of C_2 is arbitrary, and to be consistent with what we did in Chapter 6 for multidegree systems, we choose it to make $\psi_{max} = 1$. Therefore, $C_2 = 1$ for all modes.

We now have an infinite series of mode shapes, each with its associated frequency. Equation (7.2.14) tells us that the first-mode shape is a half sine wave, and Eq. (7.2.13) gives us its frequency $\omega_1 = \pi^2 (EI/ml^4)^{1/2}$. The second-mode shape is a complete sine wave at four times the first-mode frequency; the third is $1\frac{1}{2}$ sine waves at nine times the first-mode frequency; etc. We know from Fourier series that these sine functions are orthogonal over the interval $0 < x < l$, and that any arbitrary beam displacement $v(x)$ can be expressed as an infinite series:

$$v(x) = \sum_{n=1}^{\infty} q_n \sin (n\pi x/l) \qquad (7.2.15)$$

The orthogonality of the modes is not peculiar to the simply supported beam, and we now explore it in greater detail.

7.3 MODAL ORTHOGONALITY

In Sec. 7.1, we had the deformation equation

$$d^2[EI\psi'']/dx^2 - \omega^2 m\psi = 0 \qquad (7.1.12)$$

in which m and EI may vary along the length of the beam. For convenience, we take the beam to be without interior supports. End supports may or may not exist, but in any case, they do not contribute to the strain energy of the system. The existence of interior supports would not invalidate the conclusions, but they would complicate the derivation.

For mode r,

$$d^2[EI\psi_r'']/dx^2 = \omega_r^2 m\psi_r \qquad (7.3.1)$$

We multiply both sides of Eq. (7.3.1) by ψ_s and integrate from 0 to l to get

$$\int_0^l \psi_s[d^2(EI\psi_r'')/dx^2]\,dx = \omega_r^2 \int_0^l m\psi_s\psi_r\,dx \qquad (7.3.2)$$

Writing the corresponding equation for mode s, multiplying both sides by ψ_r, and integrating from 0 to l gives

$$\int_0^l \psi_r[d^2(EI\psi_s'')/dx^2]\,dx = \omega_s^2 \int_0^l m\psi_r\psi_s\,dx \qquad (7.3.3)$$

We subtract Eq. (7.3.3) from Eq. (7.3.2) to get

$$\int_0^l \psi_s[d^2(EI\psi_r'')/dx^2]\,dx - \int_0^l \psi_r[d^2(EI\psi_s'')/dx^2]\,dx$$

$$= (\omega_r^2 - \omega_s^2) \int_0^l m\psi_r\psi_s\,dx \qquad (7.3.4)$$

We evaluate the integrals on the left side of Eq. (7.3.4) by parts ($\int y\,dz = yz - \int z\,dy$). For the first integral, let

$$\begin{aligned} y &= \psi_s & dz &= [d^2(EI\psi_r'')/dx^2]\,dx \\ dy &= \psi_s'\,dx & z &= d(EI\psi_r'')/dx \end{aligned} \qquad (7.3.5)$$

Then

$$\int_0^l \psi_s[d^2(EI\psi_r'')/dx^2]\,dx = [\psi_s\,d(EI\psi_r'')/dx]_{x=0}^{x=l}$$

$$- \int_0^l [d(EI\psi_r'')/dx]\psi_s'\,dx \qquad (7.3.6)$$

Using parts again for the last integral in Eq. (7.3.6), with

$$\begin{aligned} y &= \psi_s' & dz &= [d(EI\psi_r'')/dx]\,dx \\ dy &= \psi_s''\,dx & z &= EI\psi_r'' \end{aligned} \qquad (7.3.7)$$

we finally get

$$\int_0^l \psi_s[d^2(EI\psi_r'')/dx^2]\,dx = [\psi_s\,d(EI\psi_r'')/dx]_{x=0}^{x=l}$$

$$- [\psi_s'EI\psi_r'']_{x=0}^{x=l} + \int_0^l EI\psi_r''\psi_s''\,dx \qquad (7.3.8)$$

The supports at the ends of the beam, if they exist, either prevent displacement completely or offer no resistance to displacement, and they either prevent rotation completely or offer no resistance to rotation. Therefore, either $\psi = 0$ or $d(EI\psi'')/dx = 0$ at each support and either $\psi' = 0$ or $EI\psi'' = 0$ at each support. The terms in square brackets on the right side of Eq. (7.3.8) are, therefore, zero, leaving simply

$$\int_0^l \psi_s[d^2(EI\psi_r'')/dx^2]\,dx = \int_0^l EI\psi_r''\psi_s''\,dx \qquad (7.3.9)$$

Similar integration by parts of the second integral in Eq. (7.3.4) give us the same result as Eq. (7.3.9) except that subscripts r and s are interchanged:

$$\int_0^l \psi_r[d^2(EI\psi_s'')/dx^2]\,dx = \int_0^l EI\psi_s''\psi_r''\,dx \qquad (7.3.10)$$

Substituting Eqs. (7.3.9) and (7.3.10) into Eq. (7.3.4), we get

$$(\omega_r^2 - \omega_s^2)\int_0^l m\psi_r\psi_s\,dx = 0 \qquad (7.3.11)$$

Therefore, if $\omega_r \neq \omega_s$,

$$\int_0^l m\psi_r\psi_s\,dx = 0 \qquad (7.3.12)$$

and this, combined with Eqs. (7.3.2) and (7.3.9), leads to

$$\int_0^l EI\psi_r''\psi_s''\,dx = 0 \qquad (7.3.13)$$

Equations (7.3.12) and (7.3.13) are the orthogonality relations for the modes. If there are repeated frequencies, functions ψ still exist such that the orthogonality relations hold for all $r \neq s$, including those for which $\omega_r = \omega_s$.

7.4 MODAL EQUATIONS OF MOTION

Now we express the displacement of the beam in terms of the mode shape functions; thus,

$$v(x, t) = \sum_{s=1}^{\infty} \psi_s(x)q_s(t) \qquad (7.4.1)$$

Then

$$\partial^2 v / \partial t^2 = \sum_{s=1}^{\infty} \psi_s \ddot{q}_s$$

$$\partial^2 v / \partial x^2 = \sum_{s=1}^{\infty} \psi_s'' q_s \qquad (7.4.2)$$

$$\partial^2 (EI \, \partial^2 v / \partial x^2) / \partial x^2 = \sum_{s=1}^{\infty} [d^2(EI\psi_s'')/dx^2] q_s$$

We put Eqs. (7.4.2) into the partial differential equation of forced vibration, Eq. (7.1.4), to get

$$\sum_{s=1}^{\infty} (m\psi_s \ddot{q}_s) + \sum_{s=1}^{\infty} ([d^2(EI\psi_s'')/dx^2] q_s) = f(x, t) \qquad (7.4.3)$$

Now we multiply both sides of Eq. (7.4.3) by ψ_r, term by term, and integrate under the summation signs to get

$$\sum_{s=1}^{\infty} \int_0^l m\psi_s \psi_r \, dx \, \ddot{q}_s + \sum_{s=1}^{\infty} \int_0^l [d^2(EI\psi_s'')/dx^2] \psi_r \, dx \, q_s = \int_0^l f\psi_r \, dx \qquad (7.4.4)$$

By virtue of Eqs. (7.3.12) and (7.3.13), all terms in the summations in Eq. (7.4.4) vanish except those for which $r = s$, leaving

$$\int_0^l m\psi_r^2 \, dx \, \ddot{q}_r + \int_0^l [d^2(EI\psi_r'')/dx^2] \psi_r \, dx \, q_r = \int_0^l f\psi_r \, dx \qquad (7.4.5)$$

Equation (7.3.9), with $s = r$, reduces Eq. (7.4.5) to

$$\int_0^l m\psi_r^2 \, dx \, \ddot{q}_r + \int_0^l EI(\psi_r'')^2 \, dx \, q_r = \int_0^l f\psi_r \, dx \qquad (7.4.6)$$

or

$$\ddot{q}_r + \frac{\int_0^l EI(\psi_r'')^2 \, dx}{\int_0^l m\psi_r^2 \, dx} q_r = \frac{\int_0^l f\psi_r \, dx}{\int_0^l m\psi_r^2 \, dx} \qquad (7.4.7)$$

or

$$\ddot{q}_r + \omega_r^2 q_r = a_r(t) \qquad (7.4.8)$$

where $a_r(t)$ is the generalized acceleration for mode r,

$$a_r(t) = \frac{\displaystyle\int_0^l f(x, t)\psi_r(x)\, dx}{\displaystyle\int_0^l m(x)\psi_r^2(x)\, dx} \tag{7.4.9}$$

As in the multidegree system, we have an effective mass, effective stiffness, and effective force for each mode, but this time they are integrals instead of finite sums. They are

Effective mass: $$m_r^* = \int_0^l m\psi_r^2\, dx \tag{7.4.10}$$

Effective stiffness: $$k_r^* = \int_0^l EI(\psi_r'')^2\, dx \tag{7.4.11}$$

Effective force: $$f_r^* = \int_0^l f\psi_r\, dx \tag{7.4.12}$$

7.5 DAMPING

Damping remains an embarrassment, perhaps even more elusive for the continuous system than for the multidegree system. If we wanted to incorporate a distributed viscous damping force into the partial differential equation of motion, Eq. (7.1.4), by means of a damping coefficient c, then if c were proportional to m, the analysis of Sec. 7.4 would yield valid modal equations of motion independent of each other. Otherwise, most likely the modal equations of motion would remain coupled in the velocity terms. In any event, while one could easily assign distributed damping mathematically, it would be very difficult to determine a valid distributed damping coefficient from the engineering properties of the system. About the only practical means of accommodating damping in a continuous system is to set it aside, find the undamped modes and frequencies, and then assign a fraction of critical damping to each mode.

7.6 THE RAYLEIGH QUOTIENT

Equation (7.4.7) implicitly contained the relation

$$\omega_r^2 = \frac{\displaystyle\int_0^l EI(\psi_r'')^2\, dx}{\displaystyle\int_0^l m\psi_r^2\, dx} \tag{7.6.1}$$

The right side of Eq. (7.6.1) is the Rayleigh quotient for a continuous system. Equation (7.6.1) corresponds to the frequency equation obtained for multidegree systems in Eq. (6.3.3).

7.7 RESPONSE TO SUPPORT MOTION

If the system is driven by motion of the supports instead of a driving force, the derivation is only slightly different. Let $v_g(t)$ be the displacement of the supports and let v be the displacement of the beam relative to the supports. The total displacement of the beam is then $v + v_g$. The equation of motion for the element in Fig. 7.1 then is

$$(\partial V/\partial x)\, \delta x = m[\partial^2(v + v_g)/\partial t^2]\, \delta x \tag{7.7.1}$$

which leads to the partial differential equation of motion

$$m(\partial^2 v/\partial t^2) + \partial^2(EI\partial^2 v/\partial x^2)/\partial x^2 = -m(\partial^2 v_g/\partial t^2) \tag{7.7.2}$$

Following this change through the rest of the derivation leads to the modal equation:

$$\int_0^l m\psi_r^2\, dx\, \ddot{q}_r + \int_0^l EI(\psi_r'')^2\, dx\, q_r = -\int_0^l m\psi_r\, dx\, \ddot{v}_g(t) \tag{7.7.3}$$

or

$$\ddot{q}_r + \omega_r^2 q_r = -\Gamma_r \ddot{v}_g t \tag{7.7.4}$$

where Γ_r is the modal participation factor:

$$\Gamma_r = \frac{\displaystyle\int_0^l m(x)\psi_r(x)\, dx}{\displaystyle\int_0^l m(x)\psi_r^2(x)\, dx} \tag{7.7.5}$$

7.8 BASE-SHEAR EQUIVALENT MASS

The concept of a base-shear equivalent mass that we explored in Sec. 5.14 for MDF systems pertains to continuous systems as well. If such a system, subjected to a base translation, has a displacement $q_r(t)$ in mode r, the displacement, from Eq. (7.1.6), is

$$v(x, t) = \psi_r(x)q_r(t) \tag{7.8.1}$$

The restoring force per unit length, from Fig. 7.1 and Eqs. (7.1.1) to (7.1.3), is

$$R(x, t) = \partial^2(EI\, \partial^2 v / \partial x^2) / \partial x^2 \qquad (7.8.2)$$

which, with Eqs. (7.1.7) and (7.1.12), can be reduced to

$$R(x, t) = \omega_r^2 m(x) \psi_r(x) q_r(t) \qquad (7.8.3)$$

To get the base shear, that is, the total force transmitted to the system by the supports, we integrate R over the length of the beam:

$$\text{Base shear} = \omega_r^2 \int_0^l m(x) \psi_r(x)\, dx\, q_r(t) \qquad (7.8.4)$$

The differential equation for q_r is

$$\ddot{q}_r + \omega_r^2 q_r = -\Gamma_r \ddot{v}_g t \qquad (7.7.4)$$

If we had a SDF spring–mass system of the same natural frequency and a mass m_r^v, subjected to the same base translation, the force transmitted to the system by the support would be

$$\text{Base shear} = \omega_r^2 m_r^v v(t) \qquad (7.8.5)$$

and the differential equation for $v(t)$ would be

$$\ddot{v} + \omega_r^2 v = -\ddot{v}_g(t) \qquad (7.8.6)$$

Hence, the base shear for mode r of the continuous system would be equal to the base shear of the SDF system if the mass of the SDF system were

$$m_r^v = \Gamma_r \int_0^l m(x) \psi_r(x)\, dx \qquad (7.8.7)$$

which, with Eq. (7.7.5), becomes

$$\text{Base-shear equivalent mass} = m_r^v = \frac{\left(\int_0^l m(x) \psi_r(x)\, dx \right)^2}{\int_0^l m(x) \psi_r^2(x)\, dx} \qquad (7.8.8)$$

As for the MDF system, the sum of the base-shear equivalent masses of all modes of the continuous system is equal to the total mass of the system:

$$\sum_{r=1}^{\infty} m_r^v = \int_0^l m(x)\, dx \qquad (7.8.9)$$

We leave the proof to the reader.

In passing, we point out that a completely unsupported system is not excited by base translation, and that for such a system, the modal participation factors and base-shear equivalent masses are irrelevant.

7.9 EXAMPLE: UNIFORM CANTILEVER BEAM

We now consider the case of a uniform cantilever beam, which is somewhat less trivial than the simply supported beam example of Sec. 7.2. We will derive the first few frequencies and mode shapes and determine the effective masses, effective stiffnesses, modal participation factors, and modal equations of motion.

Equation (7.2.5) pertains to the uniform cantilever beam as well as to the simply supported beam. For a cantilever fixed at end $x = 0$, the displacement and slope at the fixed end are zero:

$$v(0) = 0 \rightarrow \psi(0) = 0 \rightarrow C_1 + C_3 = 0 \rightarrow C_3 = -C_1 \quad (7.9.1)$$

$$v'(0) = 0 \rightarrow \psi'(0) = 0 \rightarrow C_2 + C_4 = 0 \rightarrow C_4 = -C_2 \quad (7.9.2)$$

and the moment and shear at the free end are zero:

$$\mathbf{M}(l) = -EI\psi''(l) = 0$$
$$\rightarrow \lambda^2(-C_1 \cos \lambda l - C_2 \sin \lambda l + C_3 \cosh \lambda l + C_4 \sinh \lambda l) = 0$$
$$(7.9.3)$$

$$\mathbf{V}(l) = -EI\psi'''(l) = 0$$
$$\rightarrow \lambda^3(C_1 \sin \lambda l - C_2 \cos \lambda l + C_3 \sinh \lambda l + C_4 \cosh \lambda l) = 0$$
$$(7.9.4)$$

We put Eqs. (7.9.1) and (7.9.2) into Eqs. (7.9.3) and (7.9.4) to get

$$(\cos \lambda l + \cosh \lambda l)C_1 + (\sin \lambda l + \sinh \lambda l)C_2 = 0 \quad (7.9.5)$$

$$(\sin \lambda l - \sinh \lambda l)C_1 - (\cos \lambda l + \cosh \lambda l)C_2 = 0 \quad (7.9.6)$$

If C_1 and C_2 were both zero, we would have the trivial case of no vibration at all. For either of them to be nonzero, the matrix of coefficients of C_1 and C_2 in Eqs. (7.9.5) and (7.9.6) must be singular; hence, its determinant must be zero. Thus,

$$\begin{vmatrix} (\cos \lambda l + \cosh \lambda l) & (\sin \lambda l + \sinh \lambda l) \\ (\sin \lambda l - \sinh \lambda l) & (-\cos \lambda l - \cosh \lambda l) \end{vmatrix} = 0 \quad (7.9.7)$$

which reduces to

$$\cos \lambda l \cosh \lambda l + 1 = 0 \quad (7.9.8)$$

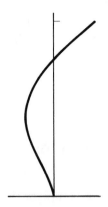

Mode 1	Mode 2
$\lambda l = 1.875$	$\lambda l = 4.694$
$\omega_1 = 3.516(EI/ml^4)^{1/2}$	$\omega_2 = 22.03(EI/ml^4)^{1/2}$
$m_1^* = 0.250ml$	$m_2^* = 0.250ml$
$k_1^* = 3.09EI/l^3$	$k_2^* = 121EI/l^3$
$\Gamma_1 = 1.566$	$\Gamma_2 = -0.867$
$m_1^v = 0.613ml$	$m_2^v = 0.189ml$

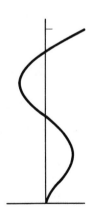
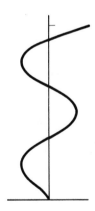

Mode 3	Mode 4
$\lambda l = 7.855$	$\lambda l = 11.00$
$\omega_3 = 61.70(EI/ml^4)^{1/2}$	$\omega_4 = 120.9(EI/ml^4)^{1/2}$
$m_3^* = 0.250ml$	$m_4^* = 0.250ml$
$k_3^* = 952EI/l^3$	$k_4^* = 3654EI/l^3$
$\Gamma_3 = 0.509$	$\Gamma_4 = -0.364$
$m_3^v = 0.065ml$	$m_4^v = 0.033ml$

Figure 7.2 Modes of uniform cantilever beam.

We solve Eq. (7.9.8) numerically to get λl, and then get ω from Eq. (7.2.3). We arbitrarily assign unit value to C_1, and then find C_2 from either Eq. (7.9.5) or (7.9.6). Then Eqs. (7.9.1) and (7.9.2) give us C_3 and C_4, the negatives of C_1 and C_2. Finally, we scale the coefficients C_i to make the maximum value of $\psi = 1$. The modal participation factor Γ requires the evaluation of two integrals in Eq. (7.7.5), which we do numerically using Simpson's rule. The denominator of Eq. (7.7.5) is the effective mass as defined by Eq. (7.4.10), and the effective stiffness is ω_r^2 times the effective mass. The two integrals in Eq. (7.7.5) also determine the base shear equivalent mass, from Eq. (7.8.8). Figure 7.2 shows the first few mode shapes, frequencies, and mode constants.

A word of caution: if more than the first few modes were needed, error control could become a problem because the mode shapes involve small differences between large numbers.

In Sec. 1.10 we solved a comparable problem approximately, taking the static deflected shape to be the first mode shape. Comparing that solution with the results shown in Fig. 7.2, we find that the approximate coefficients in Eq. (1.10.24) differ from the exact coefficients in Eq. (7.7.4) by less than 1 percent.

PROBLEMS

7.1 Find the first three modes of a uniform beam fixed at one end and simply supported at the other.

7.2 Find the first three modes of a uniform beam free at both ends.

7.3 Find the first three modes of a uniform beam fixed at both ends.

7.4 Prove that Eq. (7.8.9) is valid for a vertical cantilever beam, that is, that the sum of the base-shear equivalent masses for all modes is equal to the total mass of the beam.

7.5 Prove that

$$\sum_{r=1} \Gamma_r \psi_r(x) = 1$$

for all values of x.

CHAPTER EIGHT

Finding the Modes of Continuous Systems

8.1 INTRODUCTION

The examples of Secs. 7.2 and 7.9 both involved uniform beams and we found the modes analytically, although we had to solve some equations numerically along the way. If either EI or m vary along the length of the beam, an analytical solution is rarely feasible and we must instead resort to numerical procedures. Fundamentally, we have two choices. We may either replace the continuous system by an approximate system with a finite number of degrees of freedom and get an "exact" set of modes by the procedures of Chapter 6, or we may retain the "exact" system and seek an approximate set of modes. Here we consider the latter approach—finding the approximate modes of an "exact" system.

For each mode, we have

$$d^2(EI\psi'')/dx^2 = \omega^2 m\psi \qquad (7.1.12)$$

Compare this with the flexure equation for a beam loaded with a distributed static load w:

$$d^2(EIy'')/dx^2 = w \qquad (8.1.1)$$

Thus, if the beam were loaded with an inertia load $m\omega^2\psi$, the static deflection due to that load would be ψ.

8.2 THE STODOLA PROCESS

The foregoing is reminiscent of the Stodola process we used for multidegree systems in Chapter 6. If the "load" were $m\psi$, the static "deflection" would be ψ/ω^2. If we chose an arbitrary shape $u(x)$, the shape could be expressed as an infinite series of the mode shapes; thus,

$$u(x) = \sum_{r=1}^{\infty} C_r\psi_r \qquad (8.2.1)$$

If we applied a "load" $mu(x)$ to the beam and found the "deflection" $v(x)$ from the relation

$$d^2(EIv'')/dx^2 = mu(x) \qquad (8.2.2)$$

that deflection would be

$$v(x) = \sum_{r=1}^{\infty} (C_r/\omega_r^2)\psi_r \qquad (8.2.3)$$

The first-mode component, having the smallest ω^2, would be magnified relative to the rest, and the deflected shape $v(x)$ would be closer to the first-mode shape than was $u(x)$.

   For an example, albeit both trivial and impractical, consider the uniform simply supported beam. Take the initial approximation to be a unit displacement all along the beam, that is, $u(x) = 1$. Load the beam with a load $mu(x) = m$. Integrate twice to get the bending moment, divide by EI to get the curvature, and integrate twice more to get the displacement. The result is the familiar deflection equation found in most structural handbooks:

$$v(x) = (ml^4/24EI)[(x/l) - 2(x/l)^3 + (x/l)^4] \qquad (8.2.4)$$

   Equation (8.2.4) has a constant term, the expression in the first set of parentheses, times a shape function, the expression in square brackets. To give the shape function a unit maximum value, we divide it by its maximum value, 0.3125, and multiply the constant term by the same amount to obtain

$$v(x) = (ml^4/76.8EI)[3.2(x/l) - 6.4(x/l)^3 + 3.2(x/l)^4] \qquad (8.2.5)$$

We know that the true first mode is a half sine wave, and the shape function in Eq. (8.2.5) is obviously much closer to the shape of a half sine wave than the constant $u(x) = 1$ we chose at the start. If now we repeat the process, taking u to be the expression in square brackets in Eq. (8.2.5), which has unit maximum value, load the beam with a loading $mu(x)$, integrate twice, divide by EI, and integrate twice more, we get the deflection:

$$v(x) = (ml^4/97.04EI)[3.14(x/l) - 5.18(x/l)^3$$
$$+ 2.59(x/l)^5 - 0.74(x/l)^7 + 0.18(x/l)^8] \qquad (8.2.6)$$

The true first mode would give us

$$v(x) = (1/\omega_1^2)\psi_1 = (ml^4/\pi^4 EI) \sin(\pi x/l) \qquad (8.2.7)$$

which expressed as a polynomial series with numerical coefficients is

$$v(x) = (ml^4/97.41EI)[3.14(x/l) - 5.17(x/l)^3$$
$$+ 2.55(x/l)^5 - 0.60(x/l)^7 + 0.08(x/l)^9 - \dots] \qquad (8.2.8)$$

Comparing Eqs. (8.2.6) and (8.2.8) term by term, we see that in only two cycles, the approximate shape function in Eq. (8.2.6) is quite close to the true half sine-wave mode shape in Eq. (8.2.8), and the constant term gives the frequency within a fraction of 1 percent of the true first-mode frequency.

Only for trivial cases such as this example would we be able to integrate formally to get the deflection $v(x)$. We now turn to one of the more practical numerical methods for accomplishing the task when the mass and stiffness vary along the beam length.

8.3 BEAM DEFLECTIONS BY THE NEWMARK METHOD

The Newmark method employs the equivalent concentrated load concept illustrated in Fig. 8.1.

Consider the shear and bending moment diagrams for a beam loaded with a distributed load, as shown in Fig. 8.1(a). Now suppose we had an identical beam with an identical load, but this time instead of applying the load directly to the beam, we interposed a series of hypothetical weightless stringers end to end, as shown in Fig. 8.1(b). The stringers would feel the distributed load, but the beam would feel only the stringer reactions. The shear diagram for the beam would be discontinuous at each stringer reaction and constant between, and the bending moment diagram would be a polygon. However, at the ends of the beam and at points between the ends of adjacent stringers the values of the shear and bending moment would be exactly the same as for the distributed loading. The stringer reactions are equivalent to the distributed load in the sense that they produce the same

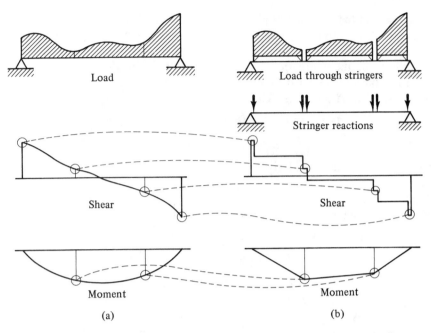

Figure 8.1 The Newmark equivalent-load concept. (a) Loading. (b) Equivalent loading.

shears and bending moments at the discrete points between the ends of adjacent stringers. Although the principle does not require that the hypothetical stringers be of equal length, computational efficiency does.

Now consider two adjacent stringers of equal length λ, loaded with a distributed load that varies parabolically over the length 2λ. Let the load intensity (force per unit length) at the stringer ends be w_0, w_1, and w_2, as shown in Fig. 8.2. The stringer reactions are then:

Left reaction:
$$R_0 = (\lambda/12)\,(3.5w_0 + 3w_1 - 0.5w_2) \qquad (8.3.1)$$

Sum of center reactions:
$$R_1 = (\lambda/12)\,(w_0 + 10w_1 + w_2) \qquad (8.3.2)$$

Right reaction:
$$R_2 = (\lambda/12)\,(-0.5w_0 + 3w_1 + 3.5w_2) \qquad (8.3.3)$$

In the Newmark method, we divide the beam into intervals of equal length, taking the interval length λ small enough to allow the load distribution over any two adjacent intervals to be approximated with suitable accuracy by a parabolic distribution. We note in passing that Eq. (8.3.2) is exact for a cubic load variation as well as for a parabolic variation.

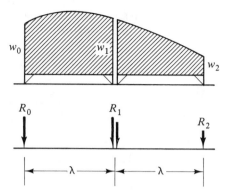

Figure 8.2 Stringer reactions.

Load, shear, and bending moment bear the same relations to each other as $\mathbf{M}/EI$, slope, and deflection. The relations are

$$\text{Load} = w \qquad\qquad \text{Curvature} = -\mathbf{M}/EI$$

$$\text{Shear, } \mathbf{V} = -\int w \, dx \qquad \text{Slope, } \theta = -\int \mathbf{M}/EI \, dx \qquad (8.3.4)$$

$$\text{Moment, } \mathbf{M} = \int \mathbf{V} \, dx \qquad \text{Deflection, } y = \int \theta \, dx$$

Thus, if we have a process for computing bending moments from loads, the same process, adjusted to account for possibly different end conditions, can be used to compute deflections from $\mathbf{M}/EI$.

8.4 NEWMARK-METHOD EXAMPLE: SIMPLY SUPPORTED BEAM

An example will clarify the Newmark method. Consider a uniform beam 20 ft long, simply supported at its ends and loaded with a triangular load of 60 kips, varying in intensity from 6 kips/ft at the left end to zero at the right. Let $E = 30,000$ ksi and $I = 500$ in^4. We divide the beam into four segments of equal length and apply the Newmark method, as shown in Table 8.1.

From the distributed load w, we compute equivalent concentrated loads as the reactions of hypothetical weightless stringers according to Eqs. (8.3.1) through (8.3.3). We then take static moments about each end

TABLE 8.1 Newmark-Method Calculations for Deflections of a Simple Beam

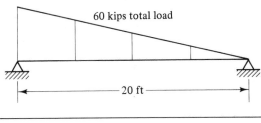

60 kips total load

20 ft

Station	0	1	2	3	4
W (kip/in)	0.500	0.375	0.250	0.125	0
Equivalent load (kips)	13.75	22.50	15.00	7.50	1.25
Reaction (kips)	40.00				20.00
Shear (kips)		26.25	3.75	−11.25	−18.75
Moment (kip-in)	0	1575	1800	1125	0
M/EI "load" (rad/in)	0	0.000105	0.000120	0.000075	0
Equivalent rotation (rad)	0.001275	0.005850	0.006900	0.004350	0.000825
End slope (rad)	0.010200				0.009000
Chord slope (rad)		0.008925	0.003075	−0.003825	−0.008175
y (in)	0	0.5355	0.7200	0.4905	0

of the beam to get the support reaction at the other end. The shear in the first panel is the left reaction minus the concentrated load at the left end, and the shear in each panel beyond the first is the shear in the preceding panel less the intervening concentrated load. The moment at the left end is zero, and the moment at the right end of each panel is the moment at the left end of the panel plus the panel length times the shear in the panel. Zero moment at the right end of the beam provides a check on the arithmetic.

We divide the moment at each panel point by EI to get the curvature at that point. Treating **M**/EI as though it were a distributed load, we convert it to equivalent concentrated "loads" (rotations), again using the hypothetical stringer concept of Fig. 8.1. We compute the end rotations the same as end reactions of a beam, and then accumulate rotations to get the chord slope in each panel, just as we computed shears. The deflection at the left end is zero, and the deflection at the right end of each panel is the deflection at the left end of the panel plus the panel length times the chord slope in the panel. Zero computed deflection at the right end provides a check on the arithmetic.

In Table 8.1, we computed the equivalent concentrated loads and reactions at the ends of the beam, but they were not really needed. We used them only to compute the shear in the first panel, which we could get equally well by other means, and to check the arithmetic. Similarly, we computed but did not need the equivalent rotations at the ends and the end slopes. We will omit them in adapting the process to the computer.

The bending moments at the panel points are exact for this example because the loading is parabolic — a degenerate parabola, to be sure. The computed deflections are exact because the M/EI diagram is a cubic and the only equivalent rotation formula that entered into the results was Eq. (8.3.2), which is exact for a cubic variation. The equivalent rotation calculations for the two ends of the beam are not exact, but their effects on the deflections are negated by the end-slope calculations.

8.5 NEWMARK-METHOD PROGRAM: SIMPLY SUPPORTED BEAM

Program 8.1 computes the bending moments and deflections for a simply supported beam by the Newmark method. Both load and EI may vary along the length of the beam. The results are "exact" if both the load and M/EI vary as polynomials of degree three or less over the length of the beam; else they are approximate.

Program 8.1: Newmark Method for Simple Beam Deflections

```
10    GOSUB 1000              ' Get and echo data
20    GOSUB 2000              ' Newmark subroutine
30    GOSUB 3000              ' Replace load with M/EI
40    GOSUB 2000              ' Newmark subroutine
50    GOSUB 4000              ' Print results
60    END

1000 F$=" #######.####    "   ' Get and echo data
1010 CLS
1020 READ N, LENGTH, E
1030 LAMBDA=LENGTH/N
1040 L12=LAMBDA/12
1050 DIM INERT(N), W(N), M(N), ECL(N), MOM(N), DEFL(N)
1060 FOR I=0 TO N: READ INERT(I): NEXT I
1070 FOR I=0 TO N: READ W(I): NEXT I
1080 PRINT "N ="; N; "   Length ="; LENGTH; "   E ="; E:
     PRINT
1090 FOR I=0 TO N
1100    PRINT USING F$; W(I);
```

```
1110    PRINT USING F$; INERT(I)
1120 NEXT I
1130 PRINT
1140 RETURN

1500 DATA 4, 240, 30000
1510 DATA 500, 500, 500, 500, 500
1520 DATA 0.5, 0.375, 0.25, 0.125, 0

2000 SHEAR=0                            ' Newmark subroutine
2010 M(0)=0
2020 M(N)=0
2030 FOR I=1 TO N-1
2040    ECL(I)=(W(I-1)+10*W(I)+W(I+1))*L12
2050    SHEAR=SHEAR+ECL(I)*(1-I/N)
2060 NEXT I
2070 FOR I=1 TO N-1
2080    M(I)=M(I-1)+SHEAR*LAMBDA
2090    SHEAR=SHEAR-ECL(I)
2100 NEXT I
2110 RETURN

3000 FOR I=0 TO N                       ' Replace load with M/EI
3010    MOM(I)=M(I)
3020    W(I)=MOM(I)/(E*INERT(I))
3030 NEXT I
3040 RETURN

4000 FOR I=0 TO N: DEFL(I)=M(I): NEXT I          ' Print
results
4010 PRINT
4020 FOR I=0 TO N
4030    PRINT USING F$; MOM(I);
4040    PRINT USING F$; DEFL(I)
4050 NEXT I
4060 RETURN
```

8.6 NEWMARK-METHOD EXAMPLE: CANTILEVER BEAM

The cantilever beam brings in a slight complication, for the end conditions for shear and moment differ from the end conditions for slope and deflection. Consider the same beam and loading, but this time let the beam be fixed at the left end and free at the right. Table 8.2 shows the Newmark-method solution.

TABLE 8.2 Newmark-Method Calculations for Deflections of a Cantilever Beam

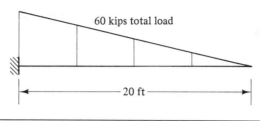

60 kips total load

20 ft

Station	0	1	2	3	4
W (kip/in)	0.500	0.375	0.250	0.125	0
Equivalent load (kips)	13.75	22.50	15.00	7.50	1.25
Reaction (kips)	60.00				0
Shear (kips)		46.25	23.75	8.75	1.25
Moment (kip-in)	−4800	−2025	−600	−75	0
M/EI "load" (rad/in)	−0.000320	−0.000135	−0.000040	−0.000005	0
Equivalent rotation (rad)	−0.007525	−0.008550	−0.002700	−0.000450	+0.000025
End slope (rad)	0				0.019200
Chord slope (rad)		0.007525	0.016075	0.018775	0.019225
y (in)	0	0.4515	1.4160	2.5425	3.6960

The equivalent loads are identical with those of Table 8.1, but this time the shear and bending moment are both zero at the right end. We therefore accumulate load effects from right to left to get the shear in each panel, and integrate shears from right to left to get the bending moments. Although we computed them, the equivalent concentrated load at the left end and the reaction at the left end serve only as a check on the arithmetic. They do not enter into the shear and moment calculations, and could just as well have been omitted.

We again compute M/EI at each panel point, treat it as a distributed load, and compute equivalent concentrated "loads" (rotations) by using the hypothetical stringer concept. Slope and deflection are both zero at the left end, so we integrate from left to right to get the deflections at the panel points.

In the simple-beam example, the bending moments and deflections were exact. Here the moments are exact, but the deflections are not. The reason lies in the formulas used for calculating equivalent concentrated rotations at the ends of the beam. Equations (8.3.1) and (8.3.3) would be exact for a parabolic distribution of M/EI "load." Equation (8.3.2) is exact also for a cubic distribution. Here the load is a linear distribution, for which all three equations are exact, thus giving the bending moments exactly. M/EI, however, varies as a cubic polynomial, and the use of Eq. (8.3.1) at the fixed end to calculate slopes injects a small error. The error is only a fraction of 1 percent, even with the coarse mesh used here.

8.7 NEWMARK-METHOD PROGRAM: CANTILEVER BEAM

Program 8.2 computes the bending moments and deflections for a cantilever beam fixed at the left end. It utilizes most of the instructions of Program 8.1, but in Program 8.2, the Newmark integration is made from right to left or from left to right, according to whether the variable INCR has a value of -1 or $+1$. The control segment first sets INCR $= -1$ and calls the Newmark subroutine, integrating from the free end to the fixed end to compute shears and bending moments, then calls for replacement of the load by $\mathbf{M}/EI$, and then sets INCR $= +1$ and calls the Newmark subroutine again, integrating from the fixed end to the free end to compute slopes and deflections. Again the load and EI may vary along the length of the beam. The results are exact if both the load and $\mathbf{M}/EI$ vary as polynomials of second degree or less; else they are approximate.

Program 8.2: Newmark Method for Cantilever Beam Deflections

```
[These instructions replace statements 20, 40, and
2000-2110 of Program 8.1 for the complete
cantilever-beam program.]
20   INCR=-1 : GOSUB 2000    ' Right to left for moments
40   INCR=1 : GOSUB 2000     ' Left to right for defl

2000 IF INCR<0 THEN N1=N ELSE N1=0   ' Newmark subroutine
2010 N2=N-N1
2020 M(N1)=0
2030 SHEAR=-INCR*(3.5*W(N1)+3*W(N1+INCR)
     -.5*W(N1+2*INCR))*L12
2040 FOR I=N1+INCR TO N2-INCR STEP INCR
2050   M(I)=M(I-INCR)+INCR*SHEAR*LAMBDA
2060   SHEAR=SHEAR-INCR*(W(I-1)+10*W(I)+W(I+1))*L12
2070 NEXT I
2080 M(N2)=M(N2-INCR)+INCR*SHEAR*LAMBDA
2090 RETURN
```

8.8 MODES BY THE STODOLA-NEWMARK PROCESS

Now let us assemble these procedures into an algorithm that will compute the modes of vibration for a cantilever beam. We can build the algorithm using a Stodola iteration that invokes the Newmark subroutine twice in each cycle, once to compute bending moments and a second time to compute deflections.

Input instructions must provide the material properties, the dimensions, and the number of segments to be used in the numerical process, which we take to be an *even* number. They must then read or compute the mass per unit length and the moment of inertia of the beam at the discrete panel points $0, 1, \ldots, N$.

To get the first mode, we begin with an arbitrary shape $u(x)$, load the beam with a loading $m(x) \cdot u(x)$, and compute the resulting deflection $v(x)$. We then divide $v(x)$ by its maximum value to get a new shape $u(x)$ with unit maximum value and repeat the process, continuing until the change produced in $u(x)$ by another cycle of iteration is negligible. We then take the latest $u(x)$ as the mode shape ψ_1.

To compute the frequency, we use the Rayleigh quotient formula, Eq. (7.6.1):

$$\omega_r^2 = \frac{\int_0^l EI(\psi_r'')^2 \, dx}{\int_0^l m\psi_r^2 \, dx} \tag{7.6.1}$$

Equation (7.3.9), with $r = s$, gives us

$$\int_0^l EI(\psi_r'')^2 \, dx = \int_0^l [d^2(EI\psi_r'')/dx^2]\psi_r \, dx \tag{8.8.1}$$

which enables us to recast Eq. (7.6.1) in a form more useful for present purposes:

$$\omega_r^2 = \frac{\int_0^l [d^2(EI\psi_r'')/dx^2]\psi_r \, dx}{\int_0^l m\psi_r^2 \, dx} \tag{8.8.2}$$

Recognizing that the final deflection v is proportional to the mode shape ψ_1, and that we computed v from the relation

$$d^2(EIv'')/dx^2 = m \cdot u \tag{8.8.3}$$

we can rewrite Eq. (8.8.2) as

$$\omega_r^2 = \frac{\int_0^l m \cdot u \cdot v \, dx}{\int_0^l m \cdot v^2 \, dx} \tag{8.8.4}$$

In the numerical process, we will have values of m, u, and v all recorded at equally spaced intervals. We can therefore evaluate the integrals readily by

Simpson's rule, which was the reason for choosing N to be an even number in the first place.

Other useful modal properties are the modal participation factor, which we compute as

$$\Gamma_r = \frac{\displaystyle\int_0^l mu\,dx}{\displaystyle\int_0^l mu^2\,dx} \tag{8.8.5}$$

and the base-shear equivalent mass:

$$m_r^v = \frac{\left[\displaystyle\int_0^l mu\,dx\right]^2}{\displaystyle\int_0^l mu^2\,dx} \tag{8.8.6}$$

To use the process for higher modes, we must sweep out the lower-mode components from the approximate mode shape $u(x)$ used in the iteration, in principle the same as we did for multidegree-of-freedom systems. However, for MDF systems, we were able to modify the dynamic matrix to sweep out all the lower modes automatically in each iteration. That luxury is unavailable here. Instead, we must sweep them out one by one in each cycle.

Any shape function u is a linear combination of the mode shapes, as expressed in Eq. (8.2.1). The modal coefficients C_r are

$$C_r = \frac{\displaystyle\int_0^l mu\psi_r\,dx}{\displaystyle\int_0^l m\psi_r^2\,dx} \tag{8.8.7}$$

For any function u, we can evaluate the coefficients C_r for the lower modes already computed, remove those modes, scale what remains to give it unit maximum value, and iterate with that. In theory, we could sweep out the lower modes once and they would be gone forever. Practice is less benign. We approximate continuous functions by their values at discrete points, we use approximations in integrating numerically, and we calculate with finite precision. We iterate with an approximation to a function $u(x)$ that is approximately orthogonal to approximate lower modes. Errors exist, and they propagate. We must therefore sweep out the lower modes in each cycle of iteration.

There is a potential pitfall in choosing the initial trial shape for a symmetrical beam with symmetrical supports, such as the simply supported beam. For such a beam, the even-numbered modes would be antisymmetrical. A symmetrical initial $u(x)$ would therefore have a zero second-mode component, and sweeping out the first mode would leave a starting vector that would theoretically cause the process to converge to the *third* mode. In fact, error might inject a second-mode component large enough to bring about ultimate convergence to the second mode, but we couldn't count on it; and even if convergence did occur, it would be painfully slow. We may avoid this difficulty by choosing an antisymmetrical initial shape for even-numbered modes.

8.9 A STODOLA PROGRAM FOR THE CANTILEVER BEAM

Program 8.3 computes a specified number of modes of a cantilever beam. It takes the initial U to be zero at the base and 1 at all other stations, and for modes higher than the first, it sweeps out lower-mode components and scales what is left to enter the Stodola iteration with an approximate shape function having unit maximum value. Likewise, it sweeps the lower-mode components out of the computed U at the end of each cycle of iteration. The mode properties are computed when the iteration for a mode is complete, the mode shape as a double-subscripted variable PSI and the other modal properties as single-subscripted variables.

Program 8.3: Stodola–Newmark Process for Modes of a Cantilever Beam

```
[Input instructions and data are for
the example of Sec. 8.11]
10    GOSUB 1000                    ' Get data
20    FOR MODE=1 TO MODES
30      U(0)=0: FOR I=1 TO N: U(I)=1: NEXT I    ' Set {U}=1
40      IF MODE>1 THEN GOSUB 4000 ' Sweep out lower modes
50      GOSUB 3000                  ' Stodola iteration
60      GOSUB 5000                  ' Get mode properties
70    NEXT MODE
80    GOSUB 6000                    ' Print results
90    END

1000 GRAV=32.174
1010 PI=3.141593
1020 I$=" ### "
1030 E$=" ##.####^^^^ "
1040 F$=" ##.#### "
1050 READ N, MODES
1060 READ HEIGHT, BASED, TOPD, BASET, TOPT
1070 READ E, DENSCON, WTLIN
1080 DIM W(N), U(N), OLDU(N), M(N), MASS(N), INERT(N)
```

```
1090 DIM PSI(N,MODES), MPSISQ(MODES),GAMMA(MODES),
     EQUIVM(MODES)
1100 DIM FREQ(MODES), PER(MODES)
1110 EPSILON=N*.000001
1120 LAMBDA=HEIGHT/N
1130 L12=LAMBDA/12
1140 FOR I=0 TO N
1150   T=TOPT+(BASET-TOPT)*((N-I)/N)^2
1160   D=BASED+(TOPD-BASED)*I/N
1170   A=PI*(D-T)*T
1180   MASS(I)=(A*DENSCON+WTLIN)/GRAV
1190   INERT(I)=(PI/64)*(D^4-(D-2*T)^4)
1200 LPRINT USING I$; I;
1210 LPRINT USING E$; D; T; A; MASS(I); INERT(I)
1220 NEXT I
1230 LPRINT
1240 RETURN

1500 DATA 40, 5
1510 DATA 900, 72, 34, 2.666667, 0.75
1520 DATA 518400, .150, 1.6

2000 IF INCR<0 THEN N1=N ELSE N1=0    ' Newmark subroutine
2010 N2=N-N1
2020 M(N1)=0
2030 SHEAR=-INCR*(3.5*W(N1)+3*W(N1+INCR)
     -.5*W(N1+2*INCR))*L12
2040 FOR I=N1+INCR TO N2-INCR STEP INCR
2050   M(I)=M(I-INCR)+INCR*SHEAR*LAMBDA
2060   SHEAR=SHEAR-INCR*(W(I-1)+10*W(I)+W(I+1))*L12
2070 NEXT I
2080 M(N2)=M(N2-INCR)+INCR*SHEAR*LAMBDA
2090 RETURN

3000 CHANGE=1+EPSILON
3010 CYCLES=0                         ' Stodola iteration
3020 WHILE CHANGE>EPSILON
3030   FOR I=0 TO N: OLDU(I)=U(I): W(I)=MASS(I)*U(I):
       NEXT I
3040   INCR=-1: GOSUB 2000           ' Compute moments
3050   FOR I=0 TO N: W(I)=M(I)/(E*INERT(I)): NEXT I
3060   INCR=1: GOSUB 2000            ' Compute deflections
3070 FOR I=0 TO N: U(I)=M(I): NEXT I
3080   GOSUB 4000                    ' Sweep out lower modes
3090   CHANGE=0
3100   FOR I=0 TO N: CHANGE=CHANGE+ABS(OLDU(I)-U(I)):
       NEXT I
```

```
3110    CYCLES=CYCLES+1
3120     PRINT MODE; CYCLES; "  CHANGE ="; CHANGE
3130  WEND
3140  PRINT
3150  RETURN


4000  FOR R=1 TO MODE-1                  ' Sweep out lower modes
4010    Z=MASS(0)*PSI(0,R)*U(0)-MASS(N)*PSI(N,R)*U(N)
4020    FOR I=1 TO N STEP 2
4030      Z=Z+4*MASS(I)*PSI(I,R)*U(I)
          +2*MASS(I+1)*PSI(I+1,R)*U(I+1)
4040    NEXT I
4050    Z=Z*LAMBDA/(3*MPSISQ(R))
4060    FOR I=0 TO N: U(I)=U(I)-Z*PSI(I,R): NEXT I
4070  NEXT R
4500  BIGU=0                                        ' Scale {U}
4510  FOR I=0 TO N
4520    IF ABS(U(I))>>ABS(BIGU) THEN BIGU=U(I)
4530  NEXT I
4540  FOR I=0 TO N: U(I)=U(I)/BIGU: NEXT I
4550  RETURN


5000  MUV=MASS(0)*U(0)*M(0)-MASS(N)*U(N)*M(N)
5010  MVSQ=MASS(0)*M(0)^2-MASS(N)*M(N)^2
5020  FOR I=1 TO N STEP 2
5030    MUV=MUV+4*MASS(I)*U(I)*M(I)
        +2*MASS(I+1)*U(I+1)*M(I+1)
5040    MVSQ=MVSQ+4*MASS(I)*M(I)^2+2*MASS(I+1)*M(I+1)^2
5050  NEXT I
5060  FREQ(MODE)=SQR(MUV/MVSQ)
5070  PER(MODE)=2*PI/FREQ(MODE)
5080  FOR I=0 TO N: PSI(I,MODE)=U(I): NEXT I
5090  Z1=MASS(0)*U(0)-MASS(N)*U(N)
5100  Z2=MASS(0)*U(0)^2-MASS(N)*U(N)^2
5110  FOR I=1 TO N STEP 2
5120    Z1=Z1+4*MASS(I)*U(I)+2*MASS(I+1)*U(I+1)
5130    Z2=Z2+4*MASS(I)*U(I)^2+2*MASS(I+1)*U(I+1)^2
5140  NEXT I
5150  MPSI=Z1*LAMBDA/3
5160  MPSISQ(MODE)=Z2*LAMBDA/3
5170  GAMMA(MODE)=MPSI/MPSISQ(MODE)
5180  EQUIVM(MODE)=MPSI^2/MPSISQ(MODE)
5190  RETURN


6000  FOR I=N TO 0 STEP -1
6010    LPRINT USING I$; I;: LPRINT "    ";
```

```
6020    FOR R=1 TO MODES: LPRINT USING F$; PSI(I,R);:
        NEXT R
6030    LPRINT
6040 NEXT I
6050 LPRINT : LPRINT "  Freq,   "
6060 LPRINT "  rad/sec ";
6070 FOR R=1 TO MODES: LPRINT USING F$; FREQ(R);:
     NEXT R
6080 LPRINT
6090 LPRINT
6100 LPRINT "  T, sec. ";
6110 FOR R=1 TO MODES: LPRINT USING F$; PER(R);: NEXT R
6120 LPRINT
6130 LPRINT : LPRINT "  Gamma   ";
6140 FOR R=1 TO MODES: LPRINT USING E$; GAMMA(R);:
     NEXT R
6150 LPRINT
6160 LPRINT : LPRINT "BSEq Mass"
6170 LPRINT "  kip/g   ";
6180 FOR R=1 TO MODES: LPRINT USING E$; EQUIVM(R);:
     NEXT R
6190 LPRINT
6200 RETURN
```

8.10 TWO APPROACHES COMPARED

Either of two approaches will yield the modes of a continuous system — we may either find an approximate solution of the "exact" partial differential equation of motion, as in this chapter, or we may approximate the continuous system by a multidegree-of-freedom system and get an "exact" solution by one of the methods of Chapter 6. Which is better, an approximate solution for the "exact" system or the "exact" solution for an approximate system? There is no universal answer, but to explore the differences, we use both approaches to a problem for which we have an analytical solution, and then compare the solutions.

Consider the uniform cantilever beam, which we solved analytically in Sec. 7.9. Program 8.3, with modifications to its input subroutine, will get the modes of the continuous system by the Newmark–Stodola method. Alternatively, we may approximate this continuous system by a system of N degrees of freedom, using N segments of equal length, lumping the mass of a full segment at each of the interior nodes and the mass of a half segment at each end, as was done for a specific case in Problem 6.6. The inertia matrix is then diagonal, with elements

$$m_{ii} = \text{(total mass)}/N \qquad i = 1, 2, \ldots, N - 1$$

$$m_{NN} = \text{(total mass)}/2N$$

We get the elements of the static flexibility matrix $[F]$ from the relation that element f_{ij} is the displacement at node i due to a unit force at node j. We evaluate the displacements by moment–area principles. The elements of the matrix are

$$f_{ji} = f_{ij} = \frac{l^3}{2EI} \frac{i^2(j - i/3)}{N^3} \qquad \begin{aligned} i &= 1, 2, \ldots, N, \\ j &= i, i + 1, \ldots, N \end{aligned} \qquad (8.10.1)$$

We may then use Program 6.1, which is the Stodola process, or another of the procedures of Chapter 6 to get the modes.

Table 8.3 compares the frequencies for the first four modes as computed by the two alternative methods. For the first mode, the continuous system yields the closer computed frequency for all values of N. For higher modes, the lumped-mass system gives a closer computed frequency for small N, but as N increases, the continuous system approaches the exact frequency more rapidly. The frequency error tends to zero approximately as N^{-2} for the lumped-mass system, and as N^{-6} for the continuous system.

TABLE 8.3   Frequencies of a Uniform Cantilever Beam Computed by Two Approaches

$$\frac{\omega}{(EI/ml^4)^{1/2}}$$

	Mode 1		Mode 2		Mode 3		Mode 4	
N	Lumped-mass	Continuous	Lumped-mass	Continuous	Lumped-mass	Continuous	Lumped-mass	Continuous
2	3.1562	3.5848	16.258	9.255				
4	3.4180	3.5183	20.091	18.954	53.202	12.874	92.73	1.30
6	3.4718	3.5163	21.109	21.946	57.557	13.620	109.47	4.98
8	3.4910	3.5160	21.502	22.034	59.279	34.186	114.29	7.88
12	3.5048	—	21.793	22.035	60.595	59.607	117.88	37.87
16	3.5098	—	21.898	22.034	61.072	61.536	119.18	91.32
24	3.5132	—	21.973	—	61.418	61.693	120.13	119.81
32	3.5144	—	22.000	—	61.540	61.696	120.47	120.83
48	3.5153	—	22.019	—	61.627	61.697	120.71	120.90
Exact	3.5160		22.034		61.697		120.90	

By either approach, the errors generated in the Stodola process for one mode propagate to higher modes. With the lumped-mass approach, we can avoid this by going to an alternative procedure, such as the frequency-equation method of Program 6.6. The price is greater computing time and having to use an interactive program. There is no reasonable way to avoid error propagation with the continuous system.

The lumped-mass approach for the example just considered was deceptively simple, for we were able to write a concise formula for the elements of the flexibility matrix. If the beam were nonuniform, we would still obtain $[F]$ from the relation that f_{ij} is the displacement at node i due to a unit force at node j; however, we would have to resort to some numerical process such as the Newmark method to determine the elements, and the difficulty of obtaining $[F]$ might override any advantages that could be attributed to the lumped-mass approach. The continuous-system analysis is no different for a nonuniform beam than for a uniform beam.

The programs for the two approaches behave differently. While the number of arithmetic operations required for one cycle of iteration varies approximately as N^2 for the lumped-mass program, the algorithm sweeps out all the lower modes automatically in the deflection calculations. Thus, the time per cycle of iteration is essentially the same for all modes. For the continuous-beam program, on the other hand, the number of arithmetic operations in the deflection calculations for each cycle of iteration varies approximately as N, but the algorithm must sweep out the lower modes one by one after each cycle, so the total number of operations per cycle of iteration varies approximately as N times the mode number.

In using the lumped-mass approach for the foregoing example, we simply allocated the mass of the adjacent half segments of the beam at each node point. There are other, more sophisticated methods, among them a consistent-mass method that considers both translation and rotation at the nodal points, resulting in a system with $2N$ degrees of freedom instead of N, and an inertia matrix $[M]$ having nonzero elements off the diagonal as well as on. The procedure is mathematically elegant, and for a uniform beam, it is not particularly difficult. If either the mass distribution or the stiffness is nonuniform, the difficulties multiply.

8.11 EXAMPLE: REINFORCED CONCRETE CHIMNEY

Here we consider an example, a reinforced concrete stack 900 ft high with an outside diameter of 34 ft at the top, tapering linearly to 72 ft at the base. The concrete wall thickness is 9 in at the top and increases parabolically to 32 in at the base. The stack is lined with a liner weighing 1.6 kips per foot of height. We take E_c to be 3600 ksi and the unit weight of concrete to be 150 lb/ft³. We consider the stack to be fixed at the base. We compute the flexural stiffness from the gross area of concrete and we ignore the stiffness of the lining, although we include its mass in the inertia properties. We will find the first five modes and then apply response spectrum methods to find the displacements, bending moments, and stresses induced by an earthquake.

To find the modal properties, we use Program 8.3, which employs the Stodola–Newmark process of Sec. 8.8. We use 40 segments in the numerical process.

We then use the displacement response spectrum of Fig. 4.3 to find the effects of the N11°W component of the Eureka, California, earthquake of December 21, 1954, as recorded in the Eureka Federal Building. Although that earthquake was destructive, it was not a great earthquake, so the results should not be interpreted as any indication of the extreme likely effects of future earthquakes. We take damping to be 5 percent of critical damping in each mode.

Figure 8.3 shows the five mode shapes and Table 8.4 gives the frequency, period, and modal participation factor for each mode as well as the spectral displacement for the Eureka earthquake.

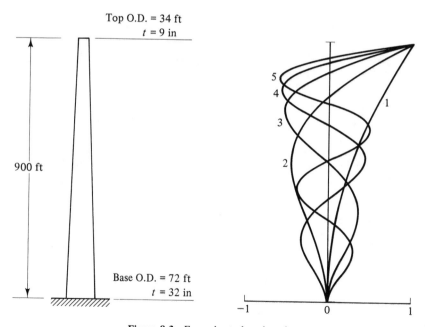

Figure 8.3 Example stack and modes.

TABLE 8.4 Modal Properties

Mode number	1	2	3	4	5
Frequency ω (rad/sec)	1.70	6.32	15.2	28.4	46.0
Period T (sec)	3.70	0.994	0.414	0.221	0.137
Modal participation factor Γ	1.97	−1.67	1.24	−0.967	0.787
Spectral displacement SD (in)	12.2	2.91	0.796	0.178	0.050
Equivalent mass/total mass	0.358	0.214	0.112	0.067	0.044

The maximum modal displacement q_r for mode r is

$$q_{r,\max} = \Gamma_r \, SD_r \qquad (8.11.1)$$

and the maximum displacement at level i for mode r is

$$x_{ir} = \psi_{ir} q_{r,\max} \qquad (8.11.2)$$

where SD_r, Γ_r, and ψ_{ir} are the spectral displacement, modal participation factor, and mode shape, respectively. To combine modes, we use the square root of the sum of the squares of the modal displacements. The algebraic sum of the modal responses is not meaningful, for spectrum methods do not reveal the times when the modal maxima occur or whether the modal responses are in phase, directly out of phase, or somewhere in between at those instants. The absolute sum would the the worst possible case if all modes were included; the root-sum-square is a more reasonable indicator of the likely maximum response.

One way of finding the bending moments and stresses for each mode would be to load the beam with a distributed inertia load of intensity $m_i \omega_r^2 \psi_{ir} q_{r,\max}$ and then use the Newmark method to get the bending moments. However, we have already done that in the final cycle of the Stodola iteration for each mode, except for a multiplying factor. When statement 5070 of Program 8.3 is reached, the value W(I)∗E∗INERT(I)∗FREQ(MODE)^2 is the bending moment at node i for a unit displacement in the mode being processed. We can record it along with the other modal properties, and then simply multiply it by the extreme modal displacement $q_{r,\max}$ to get the extreme bending moments for that mode.

Table 8.5 shows the extreme values of several earthquake-response measures. While the first mode dominates the displacement response, the higher modes contribute a great deal to the bending moments, and even more to the base shear. Indeed, the second- and third-mode base shears exceed the first-mode base shear.

TABLE 8.5 Extreme Values of Earthquake Response

Mode number	1	2	3	4	5	RSS
Displacement (in)	24.0	4.86	0.99	0.17	0.04	24.5
Base shear (kips)	1,117	2,208	1,831	856	412	3,221
Bending moment (kip-ft)	681,700	560,200	278,000	91,200	33,600	930,200
Bending stress (psi)	605	672	389	135	52	839
Gravity ± bending (psi)						+1,104
						−593

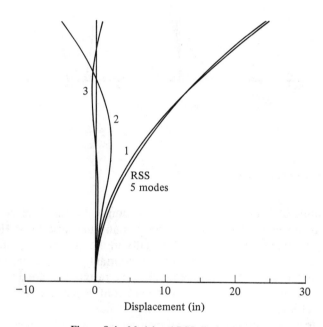

Figure 8.4 Modal and RSS displacements.

Figure 8.4 shows the displacements for each mode and the RSS combination. The mode 1 and RSS curves are almost the same, and the mode 4 and mode 5 displacements are too small to be seen in the figure.

Figure 8.5 shows the earthquake bending stresses for each mode and for the five modes combined. Here the higher modes contribute significantly to the response.

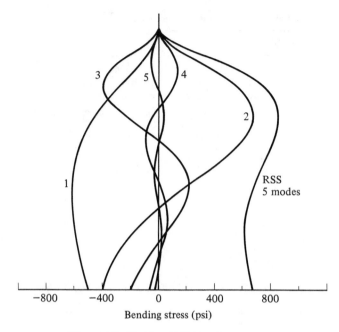

Bending stress (psi)

Figure 8.5   Modal and RSS bending stresses.

8.12 HOW MANY MODES?

In the example of Sec. 8.11, the first mode dominates the displacement response, and modes higher than the second contribute little to it. However, for bending moments and stresses, especially in the upper half of the stack, the effects of the higher modes are more important. In this case, we could reasonably assert that the first three modes give results that are adequate for engineering purposes. The other modes lend confidence, for without them, we would not know the degree of their importance.

The steadily diminishing cost of ever more powerful computing facilities has led to the use of an ever greater number of modes in dynamic analyses. Even if the higher modes are mathematically accurate, at some level, they lose their physical significance. We normally base structural analyses on the Bernoulli–Euler equations, which neglect both deformation due to shear strain and inertia due to rotation of the mass. These are truly negligible in the lower modes, but their significance increases as the mode number increases. Even the consistent-mass method, which includes rotation at the nodes among the degrees of freedom, neglects shear and rotatory inertia. The Timoshenko beam, a formulation that includes the effects of both, is a topic suited to applied-mechanics research, but too formidable for inclusion in most engineering structural dynamic analyses. No matter how elegant the analysis or how precise the computations, the accuracy of the results cannot surpass the accuracy of the mathematical model we use to represent the structure and the driving force.

PROBLEMS

8.1 A haunched rectangular cantilever beam of length l and uniform width b has a constant depth d over the right half of its length, as shown in the figure. The left half is haunched, with the depth varying parabolically from d at the midpoint to $d + h$ at the fixed end. (The vertex of the parabola is at the midpoint.) The beam is subjected to a uniformly distributed static load w per unit length.

If $l = 9$ ft, $b = 16$ in, $d = 12$ in, $h = 12$ in, $E = 3600$ ksi, and $w = 1400$ lb/ft, find the deflections at the quarter points using the Newmark method.

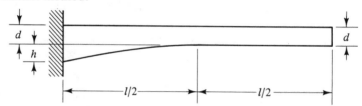

8.2 A rectangular simply supported beam has length l and a uniform width b, as shown in the figure. The depth is a uniform d over the middle third of the beam and tapers linearly at the two ends, varying from d at the third point to d_0 at the end. A uniformly distributed load w per unit length acts on the beam.

If $l = 18$ ft, $b = 12$ in, $d = 16$ in, $d_0 = 12$ in, $E = 3600$ ksi, and $w = 1.0$ kip/ft, find the deflections at the third points using the Newmark method.

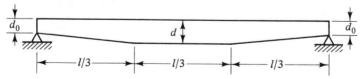

8.3 A tapered square aluminum tine is welded to a rigid housing, as shown in the figure. If its length $= 8.25$ in, tip width and depth $= 0.25$ in, base width and depth $= 0.50$ in, $E = 11,000,000$ psi, and weight density = 0.102 lb/in^3, find the natural frequency of the first mode of vibration.

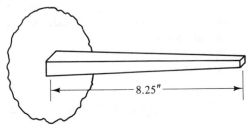

8.4 The beam of Problem 8.1 supports a dead load of 600 lb/ft in addition to its own weight of 150 lb/ft^3. Damping is negligible.
   (a) Find the first two mode shapes and frequencies.
   (b) Write the equations of motion for the first two modes of response to a uniformly distributed dynamic load $w(t)$ per unit length.

8.5 The beam of Problem 8.2 supports a dead load of 500 lb/ft in addition to its own weight of 150 lb/ft^3. Damping is negligible.
   (a) Find the first three mode shapes and frequencies.
   (b) Write the equations of motion for the first three modes of response to a single concentrated dynamic load $P(t)$ acting at the one-third point.
   (c) Write the equations of motion for the first three modes of response to a single concentrated dynamic load $P(t)$ acting at midspan.

CHAPTER NINE

Earthquake-Resistant Design

9.1 TOOLS FOR EARTHQUAKE-RESISTANT DESIGN

In the preceding eight chapters, we have considered the *elements* of structural dynamic *analysis*. Absent are many of the more sophisticated mathematical methods such as LaGrange's equations, Hamilton's principle, and LaPlace transforms. Although useful, they are not essential to an elementary treatment of the subject. Except for the numerical methods of Chapter 3, we have restricted our attention to linear systems, either undamped or viscous damped, omitting other forms of damping and energy dissipation and, except for the example of Sec. 3.18, omitting any geometric or material nonlinearities that might result from large displacements or deformation of materials beyond the linear elastic range. The omission does not imply that these topics are unimportant, but comes instead from adhering to the essential *elements* of dynamic *analysis*. This chapter touches on quite a different topic — design for earthquake resistance.

To analyze the dynamic response of a structure, we first describe its properties mathematically. If elastic behavior is assumed, which it often is, the resulting equations of motion are linear and can be solved by the methods of the preceding chapters, either as coupled equations or, with the aid of normal modes procedures, as uncoupled equations. Nonlinear-system properties preclude the use of normal modes and require solving the coupled equations. Alternatively, if the nonlinearities are not severe, we might be able to approximate the nonlinear system by one that is linear over a limited range of response. If the system is continuous, we must either approximate it by a multidegree-of-freedom system or else neglect all but a finite number of modes of the continuous system.

Buildings are unlikely to respond to dynamic excitation as truly linear systems. The behavior of such components as structural connections, stairways, and walls is particularly difficult to describe mathematically. The designer may consider curtain walls and partitions to be nonstructural components, but simply designating them as nonstructural on the plans does not make them respond that way. And unless deformations remain small, members and connections are likely to be strained beyond the linear elastic range.

If we were able to describe mathematically the properties of the entire structure and the driving force completely and unambiguously, then it would be possible, at least in theory, to calculate the response to any desired degree of precision. But even if the cost of the calculation were not prohibitive, the result would be an analysis, not a design.

The designer seeks to obtain a structure that will resist adequately any earthquakes that might affect the building during its intended useful life. The earthquake shakes the foundation beneath the building. If the building is strong enough and resilient enough, it will move along with the ground and vibrate. If it is too weak or too brittle, damage and possibly collapse will ensue. The designer wants to ensure that the completed building will have the strength and resilience necessary for satisfactory performance. The seismic building code is foremost among the tools he uses to accomplish his task. Dynamic analysis may serve to refine the design, but the building code is the essential tool.

9.2 THE DEVELOPMENT OF MODERN SEISMIC BUILDING CODES

Building codes of one form or another have existed for thousands of years. Hammurabi, king of Babylon in the eighteenth century B.C., decreed the earliest known legal requirements for buildings. The Code of Hammurabi was a code of laws, not a building code. It gave no guidance about how to build, but it imposed harsh penalties on the builder if things went wrong, even up to the death penalty if a house fell and killed the owner. Seismic building codes are relatively recent, largely because only in recent decades have we gained an adequate understanding of the nature of earthquake forces and their effects on structures.

Most seismic building codes require that structures be designed to resist specified *static* lateral forces related to the properties of the structure and the seismicity of the region. Static forces are, of course, much simpler to accommodate in the design process than dynamic forces.

Many codes do not explicitly consider vertical earthquake forces, for the reason that gravity load provisions coupled with normal factors of safety ought to be enough to accommodate the added vertical effects of the earthquake. Lateral forces are quite a different matter, for the normal state of lateral acceleration is zero, in contrast with the normal vertical acceleration of gravity.

The concept of an equivalent static lateral force could be misleading. Most engineers feel comfortable about designing for static loads. The design loads usually considered for buildings, such as floor loads and snow loads, are rarely if ever exceeded. We may be tempted to extend this line of thinking to earthquake-resistant design. However, the static lateral-design forces stipulated in most seismic building codes are not at all equal to the greatest dynamic forces likely to be exerted on the structure by an earthquake. They are "equivalent" to the dynamic forces of an earthquake only in the sense that a structure designed to resist the code forces without overstress ought to be able — if the design is carefully executed to account for stress reversals, provide adequate member ductility, and provide connections of sufficient strength and resilience — to resist minor earthquakes without damage, resist moderate earthquakes without extensive structural damage, and resist major earthquakes without collapse. The principal objective of the building code is to protect human life and public safety. To minimize economic loss is a laudable aim, but secondary. A structure competently designed by a well-qualified engineer to comply with all provisions of the code, and constructed with the best of materials and workmanship under close engineering supervision, might still be far overstressed in a major earthquake and could be expected to suffer structural as well as nonstructural damage.

Among the early regulations for the design of buildings to resist earthquakes were those developed in Italy after the Messina–Reggio earthquake of 1908. After studying timber-framed buildings that had survived that devastating earthquake with little or no damage, a commission ruled in 1909 that buildings had to be designed to withstand a lateral force of $\frac{1}{12}$ of their own weight. Three years later, this was modified to provide that the ground story must resist a lateral force of $\frac{1}{12}$ of the weight above, and the second and third stories, ⅛ of the weight above. Three-story buildings were the tallest permitted. The ruling was something of a force-equals-mass-times-acceleration approach, making no allowances for different types of construction, but at that time, there was not a great diversity of building types in the region. The provision of greater design accelerations for higher levels is consistent with structural dynamic behavior.

In Japan, after the 1923 Kanto earthquake destroyed large parts of Tokyo and Yokohama, building officials prescribed a seismic coefficient of ⅟₁₀ for all important new structures, that is, such structures had to be designed to have enough strength at any level to withstand a horizontal force of ⅟₁₀ of the weight above. In practice, many of the more conservative designers or building owners used even larger seismic coefficients.

American seismic building regulations were developed, as one might expect, largely in California. The City of San Francisco was rebuilt after the earthquake of 1906 under code provisions that stipulated a wind load of 30 lb/ft^2, which was intended to provide resistance to earthquake forces as well as wind forces. Some engineers took the mass of the structure into account in their designs, but the San Francisco code did not require it.

The State of California Division of Architecture, upon being assigned responsibility for public-school safety after the disastrous Long Beach earthquake of 1933, adopted regulations requiring that public-school buildings be designed to resist a lateral force equal to a prescribed fraction of the dead load and part of the design live load. The fraction was 10 percent for masonry buildings without frames and 2 to 5 percent for other buildings, depending on the allowable foundation loads.

In 1933, the City of Los Angeles imposed a lateral force requirement of 8 percent of the sum of dead load and half of the design live load. Los Angeles enacted a code in 1943 that related the lateral design force to the flexibility of the building — the first such code in the United States and among the first anywhere. The relationship was indirect, achieved by means of a seismic coefficient C that varied among the stories of a building. The coefficient was determined by the formula

$$C = 0.60/(N + 4.5) \qquad (9.2.1)$$

where N = number of stories above the one under consideration.

Each story had to be designed to resist a lateral shear of C times the dead load above. Thus, the total shear in a story ranged from 0.133 times the dead load above for the top story ($N = 0$) to 0.0364 times the dead load for the ground story of a 13-story building ($N = 12$), then the maximum allowable number of stories in a building. A later revision changed the formula for C and removed the 13-story height limit.

In 1947, San Francisco adopted lateral-force design requirements ranging from 3.7 to 8 percent of the design vertical load, the percentage depending on the number of stories, with variations for soil conditions. Then in
1948, the Structural Engineers Association of Northern California and the
San Francisco Section of the American Society of Civil Engineers formed a
joint committee to draft model lateral-force provisions for California building codes. By that time, the Coast and Geodetic Survey had deployed
strong-motion accelerographs and obtained a few strong-motion records at
moderately close range. Also, earthquake-response spectra had been evaluated for several strong-motion records. Meager as it was, the information
then at hand about strong-earthquake ground motion far surpassed anything
that had been available for earlier code development.

The joint committee recommended that buildings be designed for a
base shear

$$V = CW \qquad (9.2.2)$$

where W = dead load plus $\frac{1}{4}$ of design live load

   $C = 0.015/T$

   T = fundamental period of the building.

Another formula gave the distribution of this base shear along the
height of the building as lateral forces at the floor levels.

The period T was determined by the formula

$$T = 0.05h/D^{1/2} \qquad (9.2.3)$$

where h = height of the building, ft

   D = plan length of the building in the
   direction being considered, ft.

The City of San Francisco enacted a building code based on the joint committee recommendations in 1956, but with the seismic coefficient increased
one-third to

$$C = 0.02/T \qquad 0.035 \le C \le 0.075 \qquad (9.2.4)$$

Building period had now become an explicit factor in the determination of
seismic design forces for the first time.

9.3 THE UNIFORM BUILDING CODE, 1930–1973

The Uniform Building Code (UBC), which has the most widely used seismic provisions of any American building code, states its purpose thus:

The purpose of this Code is to provide minimum standards to safeguard life or limb, health, property and public welfare by regulating and controlling the design, construction, quality of materials, use and occupancy, location and maintenance of all buildings and structures within this jurisdiction and certain equipment specifically regulated herein.

The Uniform Building Code was first enacted by the International Conference of Building Officials in October, 1927. Revised editions have been published since that time at approximate three-year intervals.

The first seismic provisions published in the Uniform Building Code appeared in the 1930 edition. They stipulated lateral forces of 7.5 to 10 percent of the weight at each level, depending on foundation conditions. The weight to be considered in computing the lateral forces was dead load plus design live load, except that live load could be disregarded if it did not exceed 50 lb/ft^2. The seismic provisions were "suggested for inclusion in the Code by cities located within an area subject to earthquake shocks."

In its 1935 edition, the Uniform Building Code introduced a seismic zone map of eleven western states, with three zones, as shown in Fig. 9.1.

It gave a formula for the total lateral force at each level:

$$F = CW \qquad (9.3.1)$$

where

$$C = \begin{cases} 0.08 \text{ to } 0.16 \text{ for Zone 3} \\ 0.04 \text{ to } 0.08 \text{ for Zone 2} \\ 0.02 \text{ to } 0.04 \text{ for Zone 1} \end{cases} \begin{array}{l} \text{depending on} \\ \text{soil conditions} \end{array}$$

$$W = \text{dead load plus } \tfrac{1}{2} \text{ the design live load}$$
$$\text{(full design live load for warehouses)}$$
$$\text{at and above the level being considered}$$

$$(9.3.2)$$

These provisions prevailed through the 1946 edition.

A major change appeared in the 1949 edition of UBC. A new zone map covering all 48 contiguous states replaced the earlier one that had covered only 11 western states. The new map had four zones, including a Zone 0 that did not exist before, and was the same as its successor, the 1952 zone map shown in Fig. 9.2, except that the Zone 3 area around Puget Sound was Zone 2 in the 1949 version, and the Zone 2 area around Charleston, South Carolina, was Zone 3 in 1949.

The lateral-force formula of earlier editions, Eq. (9.3.1), was retained in 1949, but both C and W were redefined in a manner similar to the 1943 Los Angeles code.

$$C = \begin{cases} 0.60/(N + 4.5) \text{ for Zone 3} \\ 0.30/(N + 4.5) \text{ for Zone 2} \\ 0.15/(N + 4.5) \text{ for Zone 1} \end{cases} \qquad (9.3.3)$$

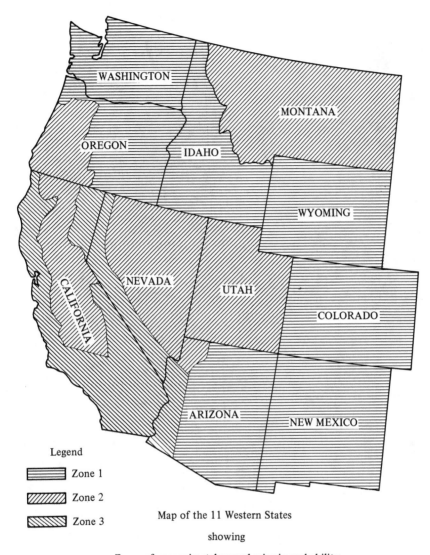

Legend

▭ Zone 1

▨ Zone 2

▧ Zone 3

Map of the 11 Western States

showing

Zones of approximately equal seismic probability

Figure 9.1   Uniform Building Code zone map, 1935. (Redrawn from the 1935 edition of the *Uniform Building Code*, copyright © 1935, with permission of the publishers, the International Conference of Building Officials.)

where N = number of stories above the story
under consideration

W = total dead-load tributary to the level being
designed (dead load plus design live load
for warehouses)

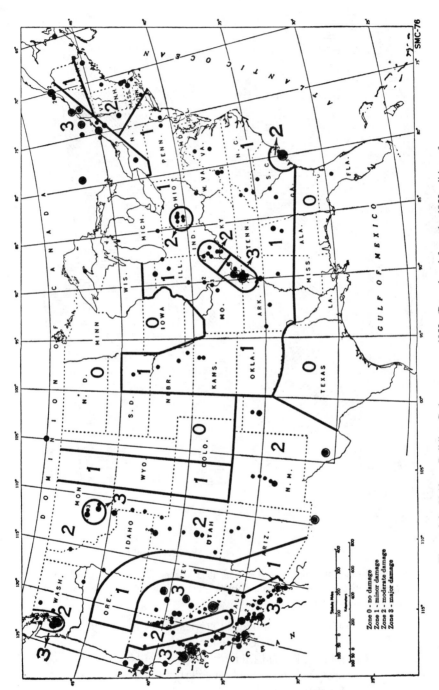

Figure 9.2 Uniform Building Code zone map, 1952. (Reproduced from the 1952 edition of the *Uniform Building Code*, copyright © 1952, with permission of the publishers, the International Conference of Building Officials.)

The base shear thus depended on the height of the building as well as its mass, an indirect recognition of the influence of building flexibility upon earthquake forces. Also, a relatively greater share of the total lateral force was allocated to the upper levels than to the lower levels.

These UBC earthquake-force design provisions remained essentially the same through the 1958 edition.

The next major revision appeared in the 1961 edition, which related lateral design forces explicitly to the building period. It stipulated a base shear

$$V = ZKCW \qquad (9.3.4)$$

where

W = total dead load (dead load plus 25 percent of live load for storage and warehouse buildings). The weight of attached machinery and movable or fixed partitions is included in the dead load for purposes of determining V.

$$Z = \begin{cases} 1 \text{ for Zone 3} \\ \frac{1}{2} \text{ for Zone 2} \\ \frac{1}{4} \text{ for Zone 1} \end{cases}$$

$K = \frac{2}{3}$ to $\frac{4}{3}$ for buildings, depending on the type of structural system employed to resist lateral forces

$C = 0.05/T^{1/3} \qquad C \le 0.10$

    where T = fundamental period (sec) of the building   (9.3.5) either determined by properly substantiated analysis or taken to be

    $T = 0.10N$ if the lateral force is resisted completely by a moment-resisting frame

    where N = number of stories, or

    $T = 0.05H/D^{1/2}$ for other buildings

    where H = height (ft) of the building

    D = length (ft) of the building parallel to the earthquake force being considered

    except that T need not be taken as less than 0.10 sec.

The base shear was distributed along the height of the building as lateral forces at the floor or roof levels:

$$F_x = Vw_x h_x \Big/ \sum_i w_i h_i \qquad (9.3.6)$$

where F_x = lateral force at level x

w_i, w_x = weight at level i, x, respectively

h_i, h_x = height above the base of level i, x, respectively

A reduction factor J was applied to these forces to determine overturning moments. The base overturning moment was

$$M = J\sum F_x h_x \tag{9.3.7}$$

where $J = 0.5/T^{2/3}$ $0.33 \le J \le 1$.

The overturning moment M_x at level x was to be determined by linear interpolation between the moment M at the base and zero at the top; thus,

$$M_x = M(H - h_x)/H \tag{9.3.8}$$

where H = height (ft) of the main portion of the building

h_x = height (ft) above the base of level x

The 1967 edition of UBC introduced changes in the base-shear distribution, stipulating a lateral force F_t at the top of the structure, which was determined as:

$$\begin{aligned} F_t &= 0 \qquad h_n/D_s \le 3 \\ F_t &= 0.004V(h_n/D_s)^2 \qquad \text{otherwise} \end{aligned} \tag{9.3.9a}$$

where h_n = height (ft) of the building

D_s = depth (ft) of lateral force resisting system

but in any case,

$$F_t \le 0.15V \tag{9.3.9b}$$

The remainder of the base shear was allocated to the various levels, including the roof, as before:

$$F_x = (V - F_t)w_x h_x \Big/ \sum wh \tag{9.3.10}$$

The base overturning moment was unchanged, except that the formula now included the extra roof force F_t:

$$M = J\left(F_t h_t + \sum_{i=1}^{n} F_i h_i\right) \tag{9.3.11}$$

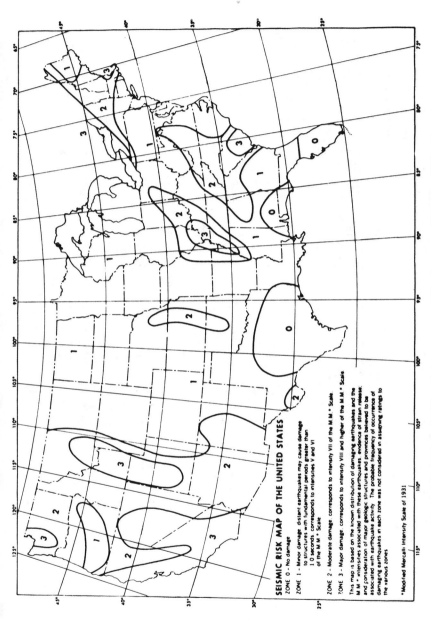

Figure 9.3 Uniform Building Code zone map, 1970. (Reproduced from the 1970 edition of the *Uniform Building Code*, copyright © 1970, with permission of the publishers, the International Conference of Building Officials.)

Overturning moments at levels above the base were computed from the static moments of the lateral forces above, with a variable reduction factor J_x, thus,

$$M_x = J_x[F_t(h_n - h_x) + \sum_{i=x}^{n} F_i(h_i - h_x)] \qquad (9.3.12)$$

where $J_x = J + (1 - J)(h_x/h_n)^3$

h_x, h_i, h_n = height (ft) above the base to level x, i, n, respectively

In the next few years, the overturning-moment reduction factor J was modified to a different formula and then eliminated entirely, in effect making $J = 1$, after exceptionally severe damage due to overturning moments occurred in Caracas in the Venezuelan earthquake of 1967.

The zone map of 1952, which had been retained through the 1967 edition of UBC, was replaced in 1970 by the zone map of Fig. 9.3. Zone maps of Hawaii and Alaska were also included.

In 1973, UBC revised the map of Fig. 9.3 by pushing the Zone 2 boundary out to the California state line, thus putting the entire state of California in Zone 3, but leaving the rest of the 1970 zone map unaltered.

9.4 THE UNIFORM BUILDING CODE, 1976–1985

The 1976 edition of UBC brought substantial changes to the earthquake regulations, and most of the 1976 regulations have been carried forward to the current (1985) edition.

A new seismic Zone 4 appeared, covering parts of California and Nevada, but the map was otherwise the same as the 1973 zone map. Then in 1982, the Zone 3 boundary in Idaho and southwestern Montana was shifted, yielding the map of Fig. 9.4, which prevails in the 1985 edition.

The formula for base shear:

$$V = ZIKCSW \qquad (9.4.1)$$

incorporates factors I and S that had not appeared before 1976. The zone coefficients are

$$Z = \begin{cases} 1 & \text{in Zone 4} \\ \frac{3}{4} & \text{in Zone 3} \\ \frac{3}{8} & \text{in Zone 2} \\ \frac{3}{16} & \text{in Zone 1} \end{cases} \qquad (9.4.2)$$

The new importance factor I, shown in Table 9.1, is intended to reflect the hazard to human life that building failure might cause.

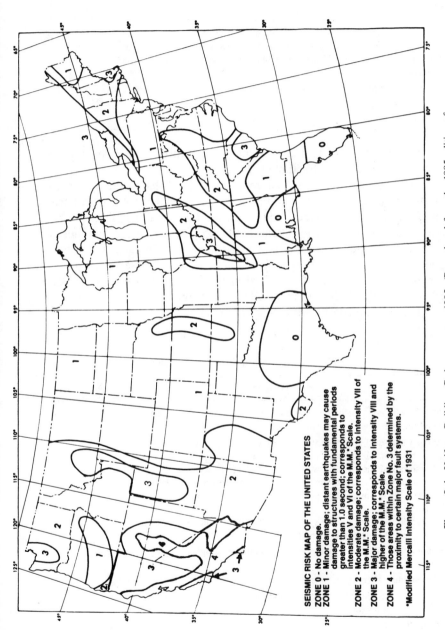

Figure 9.4 Uniform Building Code zone map, 1985. (Reproduced from the 1985 edition of the *Uniform Building Code*, copyright © 1985, with permission of the publishers, the International Conference of Building Officials.)

SEISMIC RISK MAP OF THE UNITED STATES

ZONE 0 - No damage.
ZONE 1 - Minor damage; distant earthquakes may cause damage to structures with fundamental periods greater than 1.0 second; corresponds to intensities V and VI of the M.M.* Scale.
ZONE 2 - Moderate damage; corresponds to intensity VII of the M.M.* Scale.
ZONE 3 - Major damage; corresponds to intensity VIII and higher of the M.M.* Scale.
ZONE 4 - Those areas within Zone No. 3 determined by the proximity to certain major fault systems.

*Modified Mercalli Intensity Scale of 1931

TABLE 9.1 Importance Factor *I*

Type of occupancy	*I*
Essential facilities, including hospitals and other medical facilities having surgery or emergency treatment areas, fire and police stations, and municipal government disaster operation and communication centers deemed to be vital in emergencies.	1.50
Any building where the primary occupancy is for assembly use for more than 300 persons in one room.	1.25
All others.	1.00

The structural system coefficient K, first introduced in the 1961 edition of UBC, is defined by Table 9.2.

UBC imposes a further requirement that is related to ductility but not reflected by the K factor. For all buildings in Zones 3 and 4 and all those in Zone 2 having an importance factor I greater than 1, all members in braced frames must be designed for lateral forces 25 percent greater than those determined from the code provisions. Also, connections must either be designed to develop the full capacity of the members, or else be designed for the code forces without the one-third increase in stresses normally allowed when earthquake forces are included in the design.

The seismic coefficient C is specified as

$$C = 1/15T^{1/2} \qquad C \le 0.12 \qquad (9.4.3)$$

TABLE 9.2 Structural System Coefficient *K*

Structural system	*K*
Building with box system: no complete vertical load-carrying space frame; lateral forces resisted by shear walls.	1.33
Building with dual bracing system consisting of ductile moment-resisting space frame and shear walls, designed so that: (a) Frames and shear walls resist total lateral force in accordance with their relative rigidities, considering the interaction of shear walls and frames. (b) Shear walls acting independently of space frame resist total required lateral force. (c) Ductile moment-resisting space frame has capacity to resist at least 25 percent of required lateral force.	0.80
Building with ductile moment-resisting space frame designed to resist total required lateral force.	0.67
Other building framing systems.	1.00
Elevated tanks, plus full contents, on four or more cross-braced legs and not supported by a building.	2.50
Other structures.	2.00

The code provides a formula for period:

$$T = 2\pi\left[\left(\sum_{i=1}^{n} w_i \delta_i^2\right) \Big/ \left(g \sum_{i=1}^{n} F_i \delta_i\right)\right]^{1/2} \qquad (9.4.4)$$

The forces F_i in Eq. (9.4.4) may be any rational distribution of base shear, and need not be exactly the forces obtained from code formulas; however, the displacements δ_i used in Eq. (9.4.4) must be the displacements computed for the forces used in the formula.

Alternatively, the code gives the two empirical period formulas from Eqs. (9.3.5):

$$T = \begin{cases} 0.10N \text{ for ductile frames} \\ 0.05h_n/D^{1/2} \text{ for other buildings} \end{cases} \qquad (9.4.5)$$

The site-structure resonance factor S may be computed from the ratio of the building period to the site period; thus,

$$\begin{aligned} S &= 1 + T/T_s - 0.5(T/T_s)^2 & T/T_s \le 1 \\ S &= 1.2 + 0.6T/T_s - 0.3(T/T_s)^2 & T/T_s > 1 \end{aligned} \qquad (9.4.6)$$

where T = fundamental period of the building

T_s = characteristic site period

but $S \ge 1$ in any case.

A UBC standard gives procedures for determining the site period T_s. The factor S determined from Eqs. (9.4.6) is 1.5 when the building period and site period coincide, and drops off on either side of resonance to a cut-off value of 1 when the building and site periods are widely separated.

The factor S may be taken to be 1.5 without conducting a geotechnical investigation to determine T_s, or if the building period T has been established by a properly substantiated analysis and exceeds 2.5 sec., the value of S may be determined by Eq. (9.4.6) with T_s taken to be 2.5 sec.

The 1985 edition of UBC provides the alternative of establishing S from the soil profile, according to Table 9.3.

In any case, the product CS in Eq. (9.4.1) need not be taken to exceed 0.14.

The top force F_t is a function of period:

$$\begin{aligned} F_t &= 0 & T \le 0.7 \text{ sec} \\ F_t &= 0.07TV & T > 0.7 \text{ sec} \end{aligned} \qquad (9.4.7)$$

but $F_t \le 0.25V$ in any case.

The remainder of the base shear is allocated to the various levels, including the roof, according to Eq. (9.3.10).

TABLE 9.3   Site-Structure Resonance Coefficient S Determined from Soil Profile

	Soil profile type	S
S_1	Rock of any characteristic, either shale-like or crystalline in nature (such material may be characterized by a shear wave velocity greater than 2500 ft/sec); or stiff soil conditions, where the soil depth is less than 200 ft and the soil types overlying rock are stable deposits of sands, gravels, or stiff clays.	1.0
S_2	Deep cohesionless or stiff-clay soil conditions, including sites where the soil depth exceeds 200 ft and the soil types overlying rock are stable deposits of sands, gravels, or stiff clays.	1.2
S_3	Soft to medium-stiff clays and sands, characterized by 30 ft or more of soft to medium-stiff clay with or without intervening layers of sand or other cohesionless soils.	1.5

In locations where the soil properties are not known in sufficient detail to determine the soil profile type or where the profile does not fit any of the three types, soil profile S_3 shall be used.

The calculated drift within any story due to lateral forces, unless it can be demonstrated that greater drift can be tolerated, is limited to

$$\Delta \le 0.005H \qquad K \ge 1$$
$$\Delta \le 0.005KH \qquad K < 1 \tag{9.4.8}$$

where Δ = lateral displacement within the story

H = height of the story

Equation (9.4.8) might at first glance seem to impose more stringent stiffness requirements for more ductile types of structures, but this is illusory. Equation (9.4.1) yields smaller design lateral forces for more ductile types of structures, for which $K < 1$, and the K factor in Eq. (9.4.8) merely offsets the force reduction, so that in effect the minimum stiffness requirements are the same for $K < 1$ as for $K = 1$.

The δ_i's used in Eq. (9.4.4) must be the computed deflections for the forces F_i, not the maximum deflections permitted by the drift limits of Eq. (9.4.8). To use the drift limits could lead to a period T grossly in error, especially in the lower seismic zones, for the lateral forces, Eq. (9.4.1), are proportional to the zone coefficient Z, but the drift limits, Eq. (9.4.8), are the same for all seismic zones.

9.5 STRUCTURAL DYNAMICS IN THE CODE

Many of the provisions contained in seismic building codes are either derived from or closely related to the theory of structural dynamics.

Period T

The period formula in the Uniform Building Code is

$$T = 2\pi \left[\left(\sum_{i=1}^{n} w_i \delta_i^2 \right) \Big/ \left(g \sum_{i=1}^{n} F_i \delta_i \right) \right]^{1/2} \tag{9.4.4}$$

where the F_i in the summation include the top force F_t. In Chapter 6, we had the Rayleigh quotient formula for the frequency of a mode of vibration:

$$\omega_p^2 = \frac{\{\phi\}_p^T [K] \{\phi\}_p}{\{\phi\}_p^T [M] \{\phi\}_p} \tag{6.3.3}$$

which would give a period

$$T_p = 2\pi \left(\frac{\{\phi\}_p^T [M] \{\phi\}_p}{\{\phi\}_p^T [K] \{\phi\}_p} \right)^{1/2} \tag{9.5.1}$$

The vector $\{\phi\}_p$ in Eq. (9.5.1) is the vector of displacements that would result from the application of static lateral forces $[K]\{\phi\}_p$. In Eq. (9.4.4), the deflections δ_i are the deflections that would be induced by the static forces F_i, including the top force F_t. Thus, if we take the δ_i's that would result from the code forces F_i to be an approximation of the first mode shape $\{\phi\}_1$, then

$$\{\phi\}_1^T [K] \{\phi\}_1 \text{ is equivalent to } \sum_{i=1}^{n} F_i \delta_i$$

Also,

$$\{\phi\}_1^T [M] \{\phi\}_1 \text{ is equivalent to } (1/g) \sum_{i=1}^{n} w_i \delta_i^2$$

Thus, the period formula in the code, Eq. (9.4.4), is the Rayleigh quotient formula, Eq. (9.5.1), with the δ_i's, the static deflected shape for the forces F_i, taken as an approximation of the first-mode shape $\{\phi\}_1$. The formula is quite accurate.

Seismic Coefficient C

The seismic coefficient C is related to the acceleration response spectrum. In a single-degree-of-freedom linear oscillator, the maximum base shear is

$$V = \text{PSA } m = \frac{\text{PSA}}{g} w \tag{9.5.2}$$

The Uniform Building Code gives the design base shear as

$$V = (ZIKS)CW \tag{9.5.3}$$

Thus, C corresponds to PSA expressed as a fraction of gravity.

For multidegree-of-freedom systems, the picture is more complex. In Chapter 5, we found that the maximum base shear for any mode of response to earthquake, say mode p, is

$$V_p = \text{PSA}_p \, m_p^v \tag{5.17.6}$$

where m_p^v is the base-shear equivalent mass for mode p. The sum of the base-shear equivalent masses for all modes is equal to the total mass of the structure. At any instant, the total base shear is equal to the sum of the modal base shears for all modes, but because the modes do not all reach their maxima simultaneously, the maximum total base shear is less than the sum of the modal maxima. Thus, structural dynamics gives us

$$V \leq \sum_{p=1}^{n} V_p = \sum_{p=1}^{n} \text{PSA}_p \, m_p^v \tag{9.5.4}$$

and if all of the PSA's were equal, which they are not, we would have

$$V \leq \text{PSA } m = \frac{\text{PSA}}{g} W \tag{9.5.5}$$

where W is the total weight of the structure. Thus, C and PSA have similar but by no means identical meanings. Figure 9.5 shows as functions of period, the UBC coefficient C and the pseudo-acceleration response spectra PSA/g for two earthquake components, the N11°W component recorded at Eureka, California, on December 21, 1954, and the N–S component of the Imperial Valley, California, earthquake of May 18, 1940 — the notorious El Centro earthquake. Both PSA curves are for 5 percent of critical damping.

Clearly, the code base shear is not equal to the maximum base shear due to the earthquake, and the stresses and displacements computed from a dynamic analysis of response to a strong earthquake may differ significantly from those calculated for code seismic forces.

Force Distribution

Structural dynamics gives us the restoring force at level i for mode p:

$$F_{ip} = \omega_p^2 m_i \phi_{ip} q_p \tag{9.5.6}$$

and the sum of the restoring forces for that mode is the modal base shear. Thus, for any mode, the restoring force at level i is

$$F = V m_i \phi_i \bigg/ \sum_{i=1}^{n} m_i \phi_i \tag{9.5.7}$$

Similarly, if the top force were zero, UBC would give us

$$F_x = V w_x h_x \bigg/ \sum_{i=1}^{n} w_i h_i \tag{9.3.6}$$

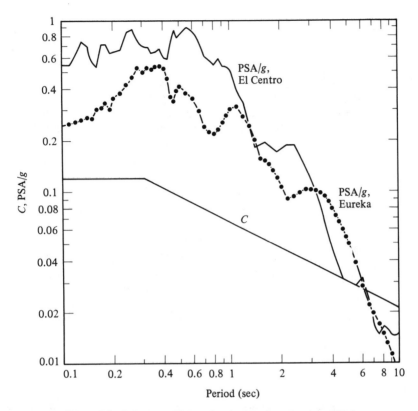

Figure 9.5 Seismic coefficient C and spectral accelerations PSA/g.

Hence, for the case $F_t = 0$, the UBC force distribution agrees with what we would derive from structural dynamics if ϕ_i were proportional to h_i, that is, if the mode shape were linear. The linear shape is not an unreasonable approximation for the first mode of a building. If the girders were very stiff compared with the columns, the shape would be concave, with greater story displacements in the lower stories. If the columns were very stiff compared with the girders, the shape would be convex, with greater story displacements in the higher stories. Real buildings are somewhere in between, and the linear shape used in the code is a convenient and not unreasonable approximation.

For higher modes, some of the restoring forces would act to the right while others acted to the left, leading to relatively greater shears in the upper stories. The top force F_t in the code also increases the story shears in the upper stories relative to the base shear. F_t, given by Eq. (9.4.7), ranges from zero for short-period structures to a maximum of one-quarter of the total base shear for long-period buildings for which the higher modes are more important.

Overturning Moment

Whereas the top force F_t reflects the influence of higher modes by increasing the shears and overturning moments in the upper stories, it increases the overturning moments even more in the lower stories. This occurs because the code forces all act in the same direction, whereas the earthquake-induced inertia forces for the higher modes act in both directions. The first mode of response is the principal source of overturning moment in the lower stories. Indeed, because the modes are orthogonal, if the first-mode shape were actually linear, the higher modes would contribute nothing at all to the overturning moment at the base, although they would affect overturning moments at higher levels. Arguments along these lines led to provisions in American codes that reduced the design overturning moment by an amount ranging from no reduction at all for short-period structures to a maximum of 67 percent of the computed base overturning moment in long-period structures. Some American codes still provide reductions as great as 55 percent of the computed overturning moment in some cases. The Uniform Building Code eliminated the overturning moment reduction factor J following the Venezuela earthquake of 1967, in which the overturning moment induced severe damage in the exterior columns of several tall buildings in Caracas.

9.6 BUILDING EXAMPLE

Let us now compute the response of the five-story example building of Fig. 6.1 to the lateral forces of the Uniform Building Code. Let us assume that the building is located in Zone 3 on a site for which a geotechnical investigation determines the site period T_s to be 2.0 sec. Assume further that the building is for office occupancy, and that it will meet the requirements for a ductile moment-resisting frame. As in Sec. 6.15, we take the story height to be 11 ft for all stories.

What follows is an analysis, not a design, for we already have the building properties. Starting a design from scratch, in order to find the base shear, we would need the weight of the building and a preliminary value of C, which is related to the yet unknown period T. An estimate of the weight might be the easiest step, for the structural framework is not a major component of total building weight and variations in the framework would affect the total weight but little.

We could start by using Eq. (9.4.5) to get a period T, find S from Eq. (9.4.6) if the site period T_s is known, or else simply take $S = 1.5$, and then compute C from Eq. (9.4.3). Then, with an estimate of the building weight W, we could compute the base shear V from Eq. (9.4.1), get the top force F_t from Eq. (9.4.7), and compute the forces F_x at the various levels

from Eq. (9.3.10). Then we could compute the story shears and the overturning moments, select the members to satisfy stress and strength requirements, compute the deflections δ_i, modify the members if necessary to meet
the drift limits of Eq. (9.4.8), and then use Eq. (9.4.4) to recompute the period T. If this value of T differed significantly from the period we used at the
start, we could repeat the process, continuing the iteration until we reached
suitable convergence.

In this case, with the building properties already known, we can use a
unit base shear to get the period from Eq. (9.4.4), for both the numerator
and denominator in Eq. (9.4.4) are proportional to the square of the base
shear. The process is still iterative because the top force F_t depends on the
period, but convergence is very rapid. Table 9.4 shows the iteration.

In the first cycle, we set the base shear $V = 1$ and assume that
$T < 0.7$ sec, for which Eq. (9.4.7) gives us $F_t = 0$. We compute the forces
F_x according to Eq. (9.3.10), add the forces above each story to get the story
shears, divide the story shears by the stiffnesses to get the story drifts, accumulate story drifts to get the displacements δ_i, and use these displacements
in Eq. (9.4.4) to compute the period T. With this period, we enter a second
cycle, again with a unit base shear, compute a new F_t, a new set of forces
F_x, and a new period T, continuing the iteration until it converges. In only
two cycles, Eq. (9.4.4) yields a period almost exactly the same as the true
first-mode period, which we found in Table 6.3 to be 0.708 sec.

TABLE 9.4   Iteration for Building Period T

Cycle		1				2		
Assumed T (sec)		< 0.7				0.7079		
F_t [Eq. (9.4.7)]		0				0.0496		
Level	F	Story shear	Story drift (in/kip)	Displ. (in/kip)	F	Story shear	Story drift (in/kip)	Displ. (in/kip)
5	0.291			0.0144	0.326			0.0150
		0.291	0.0029			0.326	0.0033	
4	0.279			0.0115	0.265			0.0117
		0.570	0.0028			0.591	0.0030	
3	0.209			0.0087	0.199			0.0088
		0.779	0.0039			0.790	0.0040	
2	0.140			0.0048	0.133			0.0048
		0.919	0.0023			0.923	0.0023	
1	0.081			0.0025	0.077			0.0025
		1.000	0.0025			1.000	0.0025	
Computed T (sec) [Eq. (9.4.4)]		0.7079				0.7078		

Now we can compute the base shear. The factors are

$Z = 0.75$ for Zone 3, from Eqs. (9.4.2)

$I = 1.0$ for office occupancy, from Table 9.1

$K = 0.67$ for a ductile moment-resisting frame, from Table 9.2

$C = 1/15T^{1/2} = 0.0792$, from Eq. (9.4.3),

$T/T_s = 0.708/2.0 = 0.354$

$S = 1 + T/T_s - 0.5(T/T_s)^2 = 1.29$, from Eq. (9.4.6),

$CS = 0.0792 * 1.29 = 0.102$; does not exceed 0.14

$W = 600$ kips

Base shear $V = ZIKCSW = 30.85$ kips, from Eq. (9.4.1)

The response is simply the response for a unit base shear, from the final cycle of iteration in Table 9.4, multiplied by the base shear. Table 9.5 shows the forces and deflections.

Table 9.6 compares these effects of code lateral forces with the effects of the Eureka earthquake computed in Chapter 6. Equation (9.4.8) limits the drift to $0.005KH$ in each story, which is 0.44 in. The earthquake response is just about at this maximum; UBC forces induce less than one-third as much. Similarly, the base shear, overturning moment, and displacements due to UBC forces are less than one-third of those due to the earthquake.

TABLE 9.5 Forces and Deflections for UBC Lateral Forces

$T = 0.708$ sec

$V = 30.85$ kips

$F_t = 0.07TV = 1.53$ kips

Level	F (kips)	Story shear (kips)	Story drift (in)	Displ. (in)
5	10.05			0.462
		10.05	0.101	
4	8.18			0.361
		18.23	0.091	
3	6.14			0.270
		24.37	0.122	
2	4.09			0.148
		28.46	0.071	
1	2.39			0.077
		30.85	0.077	
	Base overturning moment = 1232 kip-ft			

TABLE 9.6 Building Response to Earthquake and UBC Forces

	Eureka earthquake		UBC
	Mode 1	RSS 5 modes	UBC
Base shear (kips)	109.5	117.8	30.9
Base OTM (kip-ft)	4378	4381	1232
Maximum displacement (in) (flr)	1.63(5)	1.65(5)	0.46(5)
Maximum drift (in) (sty)	0.44(3)	0.45(5)	0.12(3)

This says nothing about the adequacy or inadequacy of the code provisions — only that in this particular example, with the building properties, seismic zone, K factor, and site period that were assumed, the computed responses to code forces and to one specific earthquake component differ by a factor of three.

9.7 CHIMNEY EXAMPLE

We now make a similar comparison of UBC lateral-force effects and earthquake effects for the chimney example of Sec. 8.11. Some of the UBC notation and formulas must be modified to accommodate a continuous system.
Let

z = length coordinate, with origin at the base

x = lateral displacement (a function of z)

x_t = lateral displacement at top

f = lateral force per unit length (a function of z)

w = weight per unit length (a function of z)

W = total weight

Equations (9.4.1) to (9.4.3) are unchanged. The summations in Eq. (9.4.4) change to integrals; thus,

$$T = 2\pi \left\{ \left(\int_0^l wx^2 \, dz \right) \Big/ \left[g \left(F_t x_t + \int_0^l fx \, dz \right) \right] \right\}^{1/2} \qquad (9.7.1)$$

The empirical period formula of Eq. (9.4.5) is not applicable. Equations (9.4.6) and (9.4.7) remain the same. The lateral force distribution of Eq. (9.3.10) becomes

$$f(z) = (V - F_t)wz \Big/ \int_0^l wz \, dz \qquad (9.7.2)$$

The drift limits of Eq. (9.4.8) lose their significance for a chimney, which is essentially a vertical cantilever beam. Story drift in a building corresponds to shear deformation in a chimney, which we took to be negligible in comparison with flexural deformation.

To find the period is an iterative process, much as it was for the building example of Sec. 9.6. In this case, we do not have the empirical Eq. (9.4.5) to provide a starting value of T; however, we may proceed without one. We first assign unit value to the base shear V and then arbitrarily choose a value of F_t in the range $0 \leq F_t \leq V/4$. Equation (9.4.7) tells us that F_t must lie in this range. We compute the forces $f(z)$ from Eq. (9.7.2), get the deflections $x(z)$ by the Newmark method with Program 8.2, and compute T from Eq. (9.7.1). With this T and a unit base shear V, we compute F_t from Eq. (9.4.7), and find the lateral forces $f(z)$ from Eq. (9.7.2). Then we recompute the deflections $x(z)$ by the Newmark method, Program 8.2, and finally use Eq. (9.7.1) to recompute T. Convergence is rapid; two or three cycles are usually sufficient.

For the example stack of Sec. 8.11, starting with an assumed $F_t = 0$ and using 40 segments in the numerical process, we get convergence in two cycles to a period $T = 3.68$ sec, as compared with the true first-mode period of 3.70 sec.

Now we compute the base shear; thus,

> $Z = 0.75$ for Zone 3, from Eqs. (9.4.2)
>
> $I = 1.0$, assuming the chimney is not an essential facility, from Table 9.1
>
> $K = 2.0$ for a structure other than a building, from Table 9.2
>
> $C = 1/15T^{1/2} = 0.035$, from Eq. (9.4.3)
>
> $T/T_s = 3.68/2.0 = 1.84$
>
> $S = 1.2 + 0.6(T/T_s) - 0.3(T/T_s)^2 = 1.29$, from Eq. (9.4.6),
>
> $CS = 0.035 * 1.29 = 0.045$; does not exceed 0.14
>
> $W = 34,300$ kips

Base shear $V = ZIKCSW = 2310$ kips, from Eq. (9.4.1).

In finding T, we computed the shears and bending moments and deflections for a unit base shear. We now multiply these by the computed base shear to get the response to UBC lateral forces.

Figure 9.6 shows the displacement due to the lateral forces of the Uniform Building Code, and also the first mode and RSS of five modes of displacement response to the Eureka earthquake, from Fig. 8.4. Figures 9.7 and 9.8 make similar comparisons for bending moments and bending stresses. Table 9.7 lists the extreme values of response.

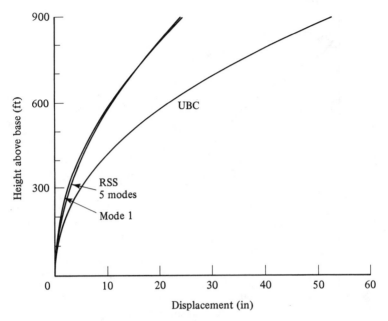

Figure 9.6 Chimney displacements due to earthquake and UBC forces.

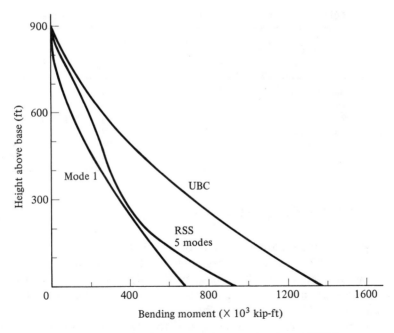

Figure 9.7 Chimney bending moments due to earthquake and UBC forces.

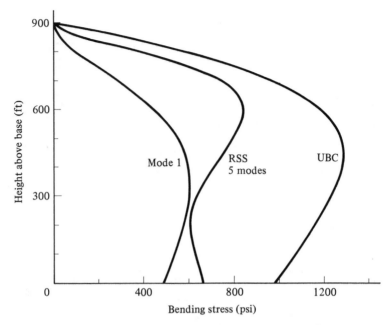

Figure 9.8 Chimney bending stresses due to earthquake and UBC forces.

TABLE 9.7 Chimney Response to Earthquake and UBC Forces

	Eureka earthquake		
	Mode 1	RSS 5 modes	UBC
Maximum displacement (in)	24.0	24.5	52.8
Base shear (kips)	1,120	3,220	2,310
Base OTM (kip-ft)	682,000	930,000	1,370,000
Maximum bending stress (psi)	605	839	1,284
Gravity ± bending (psi)		+1,104	+1,608
		−593	−975

9.8 A COMPARISON OF RESULTS

The results for the building example of Sec. 9.6 and for the chimney example of Sec. 9.7 differ in both magnitude and character. UBC lateral forces on the building led to base shear, base overturning moment, lateral displacements, and story drifts just more than one-quarter of those computed for the Eureka earthquake. UBC lateral forces on the chimney produced a base shear about two-thirds of that computed for the earthquake, stresses and bending moments about 1.5 times as great as the earthquake, and a maximum displacement more than twice as great. Yet the site conditions were the same for the two examples.

The foremost reason for the difference in magnitude is the K factor. Z and I were the same for both examples, and S turned out to be nearly the same although the ratios of T/T_s were quite different. The building was taken to meet the requirements for a ductile moment-resisting frame structure, giving it the lowest K in Table 9.2, $K = 0.67$. The chimney, being not a building structure, was assigned $K = 2$. Only elevated tanks are assigned a higher K factor.

The seismic coefficients C were quite different for the building and chimney, but are not really comparable. We see in Fig. 9.5 that both C and PSA for the first-mode period of the building, $T = 0.71$ sec, are near the highest in the chart, whereas those for the first-mode period of the chimney, $T = 3.7$ sec, are much lower.

The product $ZIKCS$ in Eq. (9.4.1) and PSA/g do have comparable meanings. For the building example, the product $ZIKCS$ was 22 percent of PSA/g at the mode-1 period and the base shear for UBC forces was 26 percent of the RSS base shear for the earthquake. Indeed, the building response to UBC lateral forces was a little more than one-fourth of the response to earthquake in all of the major response measures — base shear, displacement, story drift, and base overturning moment.

For the chimney example, $ZIKCS$ was 75 percent of PSA/g at the first-mode period and the base shear for code forces was 72 percent of the RSS base shear for the earthquake. However, other measures of response were quite different. Base overturning moment and maximum bending stress for UBC forces were nearly 1.5 times as great as for the earthquake, and the maximum displacement was more than twice as great.

Why should the ratio of code-to-earthquake response be nearly the same for all of the major response measures for the building but vary so greatly for the chimney? The answer is found principally in the influence of higher modes. The building response to the earthquake, as seen in Table 6.7, was primarily a first-mode response. For the chimney, the first mode dominated the displacement response to earthquake, but in other measures of response, the higher modes accounted for a much larger share of the total. Indeed, we see in Table 8.5 that the maximum second-mode bending stress exceeds the maximum first-mode bending stress, and both the second- and third-mode base shears surpass the first-mode base shear.

The top force F_t is intended to reflect the effects of higher modes of response in long-period structures, where the higher modes are likely to be important. F_t as a fraction of the base shear is about five times as great for the chimney as for the building. Being a fraction of the base shear, F_t does not alter the base shear at all, but it does increase the shear at higher levels. However, just as a force at the end of a cantilever beam produces more bending moment and more deflection than the same force distributed along the beam, so does the force F_t increase the bending moments, bending stresses, and displacements at all levels. The higher modes of earthquake response for the chimney affected the top displacement least, the base overturning moment more, and the base shear most. On the other hand, the UBC force F_t affects the top displacement most, the base overturning moment less, and the base shear not at all. F_t does serve to increase shears and bending moments at higher levels in the structure, which is its principal purpose. It does what it was intended to do, but it produces some side effects also.

Figure 9.9 Tied column, Olive View Hospital. (Reprinted from *Seismic Design Codes and Procedures,* by Glen V. Berg, with permission of the Earthquake Engineering Research Institute.)

9.9 SOME PRACTICAL CONSIDERATIONS

The design of earthquake-resistant structures is fully as much an art as a science, and the observed behavior of different structures in strong earthquakes has provided many valuable insights. Some of them have been translated into code provisions. With the aid of hindsight, many of them are obvious.

    1. *Design for Ductility.* Olive View Hospital in San Fernando, California, largely destroyed in the earthquake of 1971, provided an excellent illustration of the merits of ductility. The corner columns in the ground story, Fig. 9.9, were tied columns; others, Fig. 9.10, were spiral columns. A tied column fractures in a brittle manner when the ties yield or break and the concrete fractures in shear. The spiral in a spiral column confines the core concrete and inhibits brittle fracture, allowing the column to retain much of its load-carrying capacity even after it is severely distorted.

Figure 9.10 Spiral column, Olive View Hospital. (Reprinted from *Seismic Design Codes and Procedures,* by Glen V. Berg, with permission of the Earthquake Engineering Research Institute.)

2. *Provide Redundancy.* The Romanian earthquake of 1977 destroyed a computing center in Bucharest. The three-story building was supported on nine reinforced concrete columns. The ground story walls were routed around the columns, as shown in Fig. 9.11, and the walls and partitions were capped with continuous windows so that the columns provided the only resistance to lateral forces. Although the building had been designed according to a seismic building code, the columns were unable to resist the lateral forces of the earthquake and the main building collapsed, as shown in Fig. 9.12. The service towers at the two ends remained standing.

In stark contrast stands the Mt. McKinley building in Anchorage, Alaska, which was severely damaged in the Good Friday earthquake of 1964. Many spandrel beams were sheared, Fig. 9.13, and some of the columns were severely damaged, including a pier column in the end wall, Fig. 9.14, that was completely severed. The structure was highly redundant

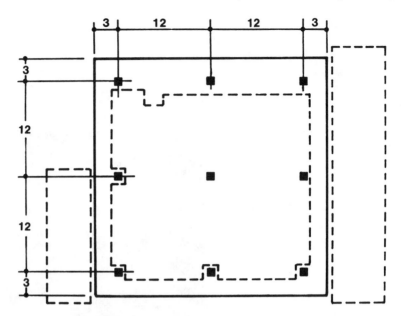

Figure 9.11 Ground-floor layout, Bucharest Computing Center. (Reprinted from *Seismic Design Codes and Procedures*, by Glen V. Berg, with permission of the Earthquake Engineering Research Institute.)

Figure 9.12 Collapsed Bucharest Computing Center. (Reprinted from *Seismic Design Codes and Procedures*, by Glen V. Berg, with permission of the Earthquake Engineering Research Institute.)

Figure 9.13 Mt. McKinley Building, Anchorage. (Reprinted from *Seismic Design Codes and Procedures,* by Glen V. Berg, with permission of the Earthquake Engineering Research Institute.)

Figure 9.14 End wall, Mt. McKinley Building. (Reprinted from *Seismic Design Codes and Procedures,* by Glen V. Berg, with permission of the Earthquake Engineering Research Institute.)

and when the capacity of one element was lost, other elements were able to pick up the load. Damage was extensive but repairable, and the repaired structure is in use today.

3. *Recognize the Principle of Relative Rigidity.* Undergoing equal distortions, a stiff component of a building will resist a greater force and incur greater damage than a flexible component. Figures 9.15 and 9.16 show two sides of a classroom in a regional college in Chimbote, Peru, damaged in the earthquake of 1972. The high walls and short windows on one side left a short effective length of columns, whereas the low walls and tall windows on the other side made the effective lengths of the columns there much greater. The short columns were stiffer than the longer ones and suffered more damage. Differences in stiffness are often caused by nonstructural components. Although the strength of nonstructural elements is usually ignored in assessing the strength of a structure, their stiffness may affect the dynamic behavior.

4. *Provide for Stress Reversals.* Earthquake forces act in all directions, and earthquake stresses will oscillate between compression and tension, between positive and negative shear. At one end of the cycle, the earthquake stresses will augment gravity load stresses, and at the other, they will counteract them. Some materials can withstand compression but not

Figure 9.15 South classroom wall, Chimbote Regional College. (Reprinted from *Seismic Design Codes and Procedures,* by Glen V. Berg, with permission of the Earthquake Engineering Research Institute.)

Figure 9.16   North classroom wall, Chimbote Regional College. (Reprinted from *Seismic Design Codes and Procedures,* by Glen V. Berg, with permission of the Earthquake Engineering Research Institute.)

tension; some beams can tolerate positive shear but not negative shear. As we saw in the building example of Sec. 9.6, the effects of an earthquake can be more severe than those due to code lateral forces, and in some cases could cause unanticipated stress reversals. One such case occurred at Koyna Dam in India, the unreinforced concrete gravity dam shown in Fig. 9.17. The lateral forces of the Indian seismic code that was used for its design, combined with gravity and the hydrostatic forces of a full reservoir, would have left the entire dam in vertical compression at all levels. The effects of the earthquake of 1967, for which the accelerations were recorded in the dam, exceeded the effects of code forces by a wide margin. The result was fracture along a horizontal plane at the upstream face due to vertical tension.

   5. *Corner Columns Are Vulnerable.* In a typical plane frame, the overturning moment due to lateral forces induces greater axial forces in the exterior columns than in the interior columns. Lateral forces due to earthquake act in both principal directions of the building, and the corner columns are exterior columns in both principal directions; hence, earthquake motion imposes greater axial forces on corner columns than on other columns. This must have contributed to the destruction of the column in Fig. 9.9.

Figure 9.17 Koyna dam. (Reprinted from *Seismic Design Codes and Procedures,* by Glen V. Berg, with permission of the Earthquake Engineering Research Institute.)

9.10 HOW EFFECTIVE IS EARTHQUAKE-RESISTANT DESIGN?

Seismic building codes replace a complex dynamic problem with a relatively simple set of static-design requirements. If the structure meets the static requirements, its response to the dynamic excitation of earthquake ought to be satisfactory. At least that is the intent. This does not imply that a strong earthquake would produce no overstress or no damage — rather, the building ought to be able to survive a moderate earthquake with insignificant damage, a major earthquake without major damage, and the strongest of earthquakes without collapse.

Several destructive earthquakes have occurred in locations where building practices were governed by building codes with modern seismic provisions. Results have been mixed. At one extreme, there have been collapses or near collapses of buildings that appeared to have complied substantially with seismic building codes, and at the other extreme, buildings not designed to resist earthquakes have survived equally severe shaking with virtually no damage at all.

Comparative evaluations have been few, but the Kern County, California, earthquake of 1952 provided a rare opportunity for comparison. After the Long Beach earthquake of 1933, which had destroyed many public-school buildings, the State of California imposed earthquake design requirements for California buildings and passed the Field Act, which charged the Division of Architecture with the responsibility of regulating the design and construction of California schools. At the time of the 1952 earthquake, there were school buildings within Kern County that had been built before adoption of the Field Act and others that had been built afterward. Those built prior to the Act had not been modified to comply with the new requirements. They were in regular use and had been well-maintained. Steinbrugge and Moran studied the effects of the earthquake and issued a comprehensive report on the damage. Table 9.8 summarizes the degree of damage they reported.

TABLE 9.8   Damage to Kern County Schools Conforming and Not Conforming to the Field Act

	Number of buildings	
Degree of damage	Conforming	Not conforming
None	11	1
Slight to minor	6	9
Moderate to considerable	1	9
Severe	0	13
Collapse	0	1

Clearly, compliance with the Field Act greatly reduced vulnerability to earthquake damage. That some buildings complying with the Act were damaged suggests that the requirements were not too severe.

Variability of material and structural properties and uncertainties about future earthquake ground motion will always stand as obstacles to the ideal building code. Structural dynamics cannot provide complete answers, but it can serve as a tool with which to improve design codes and procedures and to design structures capable of resisting earthquakes more effectively.

Answers to Problems

Chapter 1

1.1
$$m\ddot{z} + \frac{k_1 k_2}{k_1 + k_2} z = f(t)$$

1.2
$$m\ddot{z} + c\dot{z} + \frac{4k_1 k_2}{k_1 + k_2} z = f(t)$$

1.3
$$(5ma + 7m_1 a^2)\ddot{\theta} + ca\dot{\theta} + 4ka\theta = P(t)$$

1.4
$$9m_1 a\ddot{\theta} + 4c\dot{\theta} + k\theta = -4.5 m_1 \ddot{u}_g(t)$$

1.5
$$\tfrac{4}{3} m_1 a\ddot{z} + c\dot{z} + (k/4)z = P(t)$$

1.6
$$2.86 \text{ ft}$$

1.7 *Differential equation of motion:*
$$(W/g)\ddot{u} + 2T \sin \theta = 0$$

where $\theta = \tan^{-1}[u/(L/2)]$

$\quad\quad \varepsilon = (1/\cos \theta) - 1$

$\quad\quad T = T_0 + AE\varepsilon$

Conditions at instant of contact:

$$u = 0 \quad \text{and} \quad \dot{u} = v_0$$

Maximum occurs when $\dot{u}$ falls to zero.

1.8
$$\frac{152mL}{315}\ddot{z} + \frac{576EI}{5L^3}z = \frac{3L}{5}p(t)$$

1.9
$$0.2218\gamma d^2 L\ddot{z} + 0.5422(Ed^4/L^3)z = 0.375Lp(t)$$

Chapter 2

2.1
$$\omega/\omega_n < (1/2)^{1/2} \quad \text{and} \quad \omega/\omega_n > (3/2)^{1/2}$$

2.2
$$k = 1.065 \text{ kip/in}$$
$$m = 5.181 \text{ lb/(in/sec}^2)$$
$$\omega_n = 14.34 \text{ rad/sec}$$
$$\omega/\omega_n = 1.314$$
$$R_d = 1.37$$
$$M_{max} = 20.6 \text{ kip-in}$$
$$f_{max} = 11.9 \text{ ksi}$$

2.3 (a)
$$\zeta = \ln D/[\pi^2 + (\ln D)^2]^{1/2}$$

(b)
$$D = 9 \sim \zeta = 0.57$$
$$D = 11 \sim \zeta = 0.61$$

2.4 0.0325 radian rotation or 1.56 in. displacement at the damper end

2.5 Spring–mass $T_n = 2.22$ sec
Pendulum $T_n = 5.36$ sec

Chapter 3

3.2 First maximum displacement $= 0.862$ ft.; first minimum displacement $= -0.566$ ft.

3.3 $u_{max} = 5.33$ in.

Chapter 4

4.1 $u_{max} = 0.69$ in, and $f_{max} = 25$ ksi.

4.2 $u_{max} = 0.95$ in, and $f_{max} = 28.3$ ksi.

Chapter 5

5.1

$\omega_1 = 2.75$ rad/sec	$\omega_2 = 6.63$ rad/sec
$T_1 = 2.29$ sec	$T_2 = 0.95$ sec
$\{\phi\}_1 = \begin{Bmatrix} 0.414 \\ 1.000 \end{Bmatrix}$	$\{\phi\}_2 = \begin{Bmatrix} 1.000 \\ -0.414 \end{Bmatrix}$

5.2
$$q_1 = 0.707 \cos 2.75t$$
$$q_2 = 1.707 \cos 6.63t$$
$$u_1 = 0.293 \cos 2.75t + 1.707 \cos 6.63t$$
$$u_2 = 0.707 \cos 2.75t - 0.707 \cos 6.63t$$

5.3 $\omega_1 = 0.558(EI/ml^3)^{1/2}$ $\omega_2 = 2.874(EI/ml^3)^{1/2}$

$$\{\phi\}_1 = \begin{Bmatrix} 0.327 \\ 1.000 \end{Bmatrix} \qquad\qquad \{\phi\}_2 = \begin{Bmatrix} 1.000 \\ -0.655 \end{Bmatrix}$$

$$\ddot{q}_1 + (0.311EI/ml^3)q_1 = 0.270f(t)/m$$
$$\ddot{q}_2 + (8.260EI/ml^3)q_2 = 0.412f(t)/m$$

5.4   Mode p 1   2   3

ω_p	$0.684(g/l)^{1/2}$	$1.286(g/l)^{1/2}$	$1.970(g/l)^{1/2}$

$$\{\phi\}_p \qquad \begin{Bmatrix} 0.210 \\ 0.532 \\ 1.000 \end{Bmatrix} \qquad \begin{Bmatrix} -0.484 \\ -0.653 \\ 1.000 \end{Bmatrix} \qquad \begin{Bmatrix} 1.000 \\ -0.879 \\ 0.305 \end{Bmatrix}$$

5.5 $\ddot{q}_1 + (0.468g/l)q_1 = \quad 1.586\omega^2 A \sin \omega t$
$$\ddot{q}_2 + (1.653g/l)q_2 = -0.688\omega^2 A \sin \omega t$$
$$\ddot{q}_3 + (3.879g/l)q_3 = \quad 0.333\omega^2 A \sin \omega t$$

5.6 (a)
$$[M] = \begin{bmatrix} 120 & 0 & 0 \\ 0 & 120 & 0 \\ 0 & 0 & 100 \end{bmatrix} \text{kip/g}$$

$$[K] = \begin{bmatrix} 316 & -193 & 0 \\ -193 & 338 & -145 \\ 0 & -145 & 145 \end{bmatrix} \text{kip/in}$$

(b) Frequency equation:
$$-0.0250\omega^6 + 66.6\omega^4 - 40{,}900\omega^2 + 3{,}450{,}000 = 0$$

Mode p 1   2   3

ω_p	9.99 rad/sec	27.7 rad/sec	42.3 rad/sec

$$\{\phi\}_p \qquad \begin{Bmatrix} 0.555 \\ 0.821 \\ 1.000 \end{Bmatrix} \qquad \begin{Bmatrix} -0.943 \\ -0.377 \\ 1.000 \end{Bmatrix} \qquad \begin{Bmatrix} -0.800 \\ 1.000 \\ -0.452 \end{Bmatrix}$$

(c) Differential equations:
$$\ddot{q}_1 + (99.8 \text{ sec}^{-2})q_1 = (\quad 1.455 \text{ in sec}^{-2} \text{ kip}^{-1})f(t)$$
$$\ddot{q}_2 + (769 \text{ sec}^{-2})q_2 = (-0.651 \text{ in sec}^{-2} \text{ kip}^{-1})f(t)$$
$$\ddot{q}_3 + (1793 \text{ sec}^{-2})q_3 = (\quad 1.777 \text{ in sec}^{-2} \text{ kip}^{-1})f(t)$$

5.7 (a)
$$[M] = \begin{bmatrix} 120 & 0 & 0 \\ 0 & 120 & 0 \\ 0 & 0 & 100 \end{bmatrix} \text{kip/g}$$

$$[K] = \begin{bmatrix} 271 & -173 & 27.2 \\ -173 & 253 & -110 \\ 27.2 & -110 & 86.5 \end{bmatrix} \text{kip/in}$$

(b) Frequency equation:

$$-0.0250\omega^6 + 50.5\omega^4 - 20,000\omega^2 + 865,000 = 0$$

Mode p	1	2	3
ω_p	7.02 rad/sec	21.6 rad/sec	38.8 rad/sec
$\{\phi\}_p$	$\begin{Bmatrix} 0.415 \\ 0.769 \\ 1.000 \end{Bmatrix}$	$\begin{Bmatrix} -0.980 \\ -0.554 \\ 1.000 \end{Bmatrix}$	$\begin{Bmatrix} -0.949 \\ 1.000 \\ -0.450 \end{Bmatrix}$

(c) Differential equations:

$$\ddot{q}_1 + (49.2 \text{ sec}^{-2})q_1 = (1.549 \text{ in sec}^{-2} \text{ kip}^{-1})f(t)$$
$$\ddot{q}_2 + (468 \text{ sec}^{-2})q_2 = (-0.849 \text{ in sec}^{-2} \text{ kip}^{-1})f(t)$$
$$\ddot{q}_3 + (1503 \text{ sec}^{-2})q_3 = (1.555 \text{ in sec}^{-2} \text{ kip}^{-1})f(t)$$

Chapter 6

6.1 See Problem 5.6(b).

6.2 See Problem 5.6(b).

6.3 See Problem 5.7(b).

6.4

Mode p	1	2	3	4
$\{\phi\}_p$	$\begin{Bmatrix} 0.340 \\ 0.657 \\ 0.889 \\ 1.000 \end{Bmatrix}$	$\begin{Bmatrix} -0.605 \\ -0.435 \\ 0.383 \\ 1.000 \end{Bmatrix}$	$\begin{Bmatrix} 0.667 \\ -0.667 \\ -0.333 \\ 1.000 \end{Bmatrix}$	$\begin{Bmatrix} 0.207 \\ -0.548 \\ 1.000 \\ -0.979 \end{Bmatrix}$
ω_p (rad/sec)	16.06	37.82	55.58	68.43
T_p (sec)	0.391	0.166	0.113	0.092

6.5

Mode p	1	2	3	4
$\{\phi\}_p$	$\begin{Bmatrix} 0.5616 \\ 1.0000 \\ 1.0000 \\ 0.5616 \end{Bmatrix}$	$\begin{Bmatrix} 1.0000 \\ 1.0000 \\ -1.0000 \\ -1.0000 \end{Bmatrix}$	$\begin{Bmatrix} 1.0000 \\ -0.2808 \\ -0.2808 \\ 1.0000 \end{Bmatrix}$	$\begin{Bmatrix} 1.0000 \\ -0.5000 \\ 0.5000 \\ -1.0000 \end{Bmatrix}$
ω_p (rad/sec)	13.01	27.79	41.97	43.94
T_p (sec)	0.483	0.226	0.150	0.143

6.6

Mode p	1	2	3	4
$\{\phi\}_p$	$\begin{Bmatrix} 0.0959 \\ 0.3360 \\ 0.6541 \\ 1.0000 \end{Bmatrix}$	$\begin{Bmatrix} -0.4544 \\ -0.8282 \\ -0.2724 \\ 1.0000 \end{Bmatrix}$	$\begin{Bmatrix} 1.0000 \\ 0.2575 \\ -0.9174 \\ 0.8534 \end{Bmatrix}$	$\begin{Bmatrix} 1.0000 \\ -0.9853 \\ 0.6440 \\ -0.3722 \end{Bmatrix}$
ω_p (rad/sec)	29.15	171.3	453.7	790.8
T_p (sec)	0.216	0.0367	0.0138	0.0079

6.7

Mode p	1	2	3	4	5
$\{\phi\}_p$	$\begin{Bmatrix} 0.5000 \\ 0.8660 \\ 1.0000 \\ 0.8660 \\ 0.5000 \end{Bmatrix}$	$\begin{Bmatrix} -1.0000 \\ -1.0000 \\ 0.0000 \\ 1.0000 \\ 1.0000 \end{Bmatrix}$	$\begin{Bmatrix} -1.0000 \\ 0.0000 \\ 1.0000 \\ 0.0000 \\ -1.0000 \end{Bmatrix}$	$\begin{Bmatrix} -1.0000 \\ 1.0000 \\ 0.0000 \\ -1.0000 \\ 1.0000 \end{Bmatrix}$	$\begin{Bmatrix} 0.5000 \\ -0.8660 \\ 1.0000 \\ -0.8660 \\ 0.5000 \end{Bmatrix}$
ω_p (rad/sec)	15.42	61.62	137.8	238.6	341.4
T_p (sec)	0.408	0.102	0.046	0.026	0.018

Chapter 7

7.1
$$\omega_1 = 15.42(EI/ml^4)^{1/2}$$
$$\omega_2 = 49.96(EI/ml^4)^{1/2}$$
$$\omega_3 = 104.2(EI/ml^4)^{1/2}$$

Beam simply supported at $x = 0$, fixed at $x = l$.

Mode	1	2	3
λ	$3.9266/l$	$7.0686/l$	$10.2102/l$
ω	$15.42(EI/ml^4)^{1/2}$	$49.96(EI/ml^4)^{1/2}$	$104.2(EI/ml^4)^{1/2}$
C_2	0.937411	-0.935080	0.935076
C_4	0.026130	0.001126	0.000049

$$[\psi = C_2 \sin \lambda x + C_4 \sinh \lambda x]$$

7.2

Mode	1	2	3
λ	$4.7300/l$	$7.8532/l$	$10.9956/l$
ω	$22.37(EI/ml^4)^{1/2}$	$61.67(EI/ml^4)^{1/2}$	$120.9(EI/ml^4)^{1/2}$
C_1	0.500000	0.500000	0.500000
C_2	-0.491251	-0.500389	-0.499983

$$[\psi = C_1(\cos \lambda x + \cosh \lambda x) + C_2(\sin \lambda x + \sinh \lambda x)]$$

There are also two rigid-body modes, translation and rotation, both at zero frequency.

7.3
$$\omega_1 = 22.37(EI/ml^4)^{1/2}$$
$$\omega_2 = 61.67(EI/ml^4)^{1/2}$$
$$\omega_3 = 120.9 \ (EI/ml^4)^{1/2}$$

Mode	1	2	3
λ	$4.7300/l$	$7.8532/l$	$10.9956/l$
ω	$22.37(EI/ml^4)^{1/2}$	$61.67(EI/ml^4)^{1/2}$	$120.9(EI/ml^4)^{1/2}$
C_1	-0.629665	0.662592	-0.661138
C_2	0.618647	-0.663107	0.661116

$$[\psi = C_1(\cos \lambda x - \cosh \lambda x) + C_2(\sin \lambda x - \sinh \lambda x)]$$

Chapter 8

8.1 0.0052 in, 0.0267 in, 0.0624 in, and 0.1025 in.

8.2 0.153 in, 0.153 in.

8.3 $\omega_1 = 2001$ rad/sec

8.4 (a)

Mode	1	2
ω(rad/sec)	101.50	542.28
ψ		
Sta 0	0.0000	0.0000
1	0.0089	−0.0616
2	0.0436	−0.2568
3	0.1180	−0.5354
4	0.2403	−0.7274
5	0.4037	−0.6543
6	0.5927	−0.2766
7	0.7945	0.3198
8	1.0000	1.0000

(b)

$$\ddot{q}_1 + (10{,}300 \text{ sec}^{-2}) \cdot q_1 = \left(64.1\frac{\text{ft/sec}^2}{\text{kip/ft}}\right) \cdot w(t)$$

$$\ddot{q}_2 + (294{,}000 \text{ sec}^{-2}) \cdot q_2 = \left(-37.3\frac{\text{ft/sec}^2}{\text{kip/ft}}\right) \cdot w(t)$$

8.5 (a)

Mode	1	2	3
ω(rad/sec)	63.624	236.78	523.68
ψ			
Sta 0	0.0000	0.0000	0.0000
1	0.1873	−0.3829	0.5543
2	0.3650	−0.7006	0.9183
3	0.5263	−0.9121	1.0000
4	0.6667	−1.0000	0.8070
5	0.7837	−0.9658	0.4181
6	0.8764	−0.8252	−0.0538
7	0.9444	−0.6011	−0.4957
8	0.9860	−0.3166	−0.8086
9	1.0000	0.0000	−0.9215
10	0.9860	0.3166	−0.8086
11	0.9444	0.6011	−0.4957
12	0.8764	0.8252	−0.0538
13	0.7837	0.9658	0.4181
14	0.6667	1.0000	0.8070
15	0.5263	0.9121	1.0000
16	0.3650	0.7006	0.9183
17	0.1873	0.3829	0.5543
18	0.0000	0.0000	0.0000

(b)
$$\ddot{q}_1 + (4002 \text{ sec}^{-2})q_1 = \left(4.38 \frac{\text{ft/sec}^2}{\text{kip}}\right)P(t)$$

$$\ddot{q}_2 + (56{,}070 \text{ sec}^{-2})q_2 = \left(-4.26 \frac{\text{ft/sec}^2}{\text{kip}}\right)P(t)$$

$$\ddot{q}_3 + (274{,}200 \text{ sec}^{-2})q_3 = \left(-0.295 \frac{\text{ft/sec}^2}{\text{kip}}\right)P(t)$$

(c)
$$\ddot{q}_1 + (4002 \text{ sec}^{-2}) \cdot q_1 = \left(5.00 \frac{\text{ft/sec}^2}{\text{kip}}\right) \cdot P(t)$$

$$\ddot{q}_2 + (56{,}070 \text{ sec}^{-2}) \cdot q_2 = 0$$

$$\ddot{q}_3 + (274{,}200 \text{ sec}^{-2}) \cdot q_3 = \left(-5.05 \frac{\text{ft/sec}^2}{\text{kip}}\right)P(t)$$

Selected References

ALFORD, J. L., G. W. HOUSNER, and R. R. MARTEL. *Spectrum Analysis of Strong-Motion Earthquakes*. Pasadena, CA: California Institute of Technology, 1951 (revised 1964).

American Society for Testing and Materials. *Metric Practice Guide E380-74*. Philadelphia: ASTM, 1974.

ANDERSON, ARTHUR W., et al. "Lateral Forces of Earthquake and Wind," *Trans. ASCE*, 117 (1952), 716–780.

BENIOFF, H. "The Physical Evaluation of Seismic Destructiveness," *Bull. Seism. Soc. Amer.*, 24 (1934), 398–403.

BERG, GLEN V. *Seismic Design Codes and Procedures*. Berkeley, CA: Earthquake Engineering Research Institute, 1983.

BIOT, M. A. "A Mechanical Analyzer for the Prediction of Earthquake Stresses," *Bull. Seism. Soc. Amer.*, 31 (April, 1941), 151–171.

CARNAHAN, BRICE, H. A. LUTHER, and JAMES O. WILKES. *Applied Numerical Methods*. New York: Wiley, 1969.

CHOPRA, ANIL K., *Dynamics of Structures: A Primer*. Berkeley, CA: Earthquake Engineering Research Institute, 1981.

CLOUGH, RAY W., and JOSEPH PENZIEN. *Dynamics of Structures*. New York: McGraw-Hill, 1975.

CRAIG, ROY R., JR. *Structural Dynamics*. New York: Wiley, 1981.

CRANDALL, STEPHEN H. *Engineering Analysis*. New York: McGraw-Hill, 1956.

FRAZER, R. A., W. J. DUNCAN, and A. R. COLLAR. *Elementary Matrices*. Cambridge, England: The Cambridge University Press, 1955.

HAMMING, RICHARD W. *Introduction to Applied Numerical Analysis*. New York: McGraw-Hill, 1971.

HARRIS, CYRIL M. *Shock and Vibration Handbook* (3rd ed). New York: McGraw-Hill, 1987.

HOUSNER, G. W. "Strong Ground Motion," in Robert L. Wiegel (ed.), *Earthquake Engineering*. Englewood Cliffs, NJ: Prentice-Hall, 1970.

HOUSNER, G. W., and P. C. JENNINGS. *Earthquake Design Criteria*. Berkeley, CA: Earthquake Engineering Research Institute, 1982.

HUDSON, D. E. "Dynamic Tests of Full-Scale Structures," in Robert L. Wiegel (ed.), *Earthquake Engineering*. Englewood Cliffs, NJ: Prentice-Hall, 1970.

————, "Ground Motion Measurements," in Robert L. Wiegel (ed.), *Earthquake Engineering*. Englewood Cliffs, NJ: Prentice-Hall, 1970.

————, *Reading and Interpreting Strong Motion Accelerograms*. Berkeley, CA: Earthquake Engineering Research Institute, 1979.

International Conference of Building Officials, *Uniform Building Code, 1985 Edition*. Whittier, CA: ICBO, 1985.

KARMAN, THEODORE V., and MAURICE A. BIOT. *Mathematical Methods in Engineering*. New York: McGraw-Hill, 1940.

KREYZIG, ERWIN. *Advanced Engineering Mathematics*. New York: Wiley, 1962.

MEIROVITCH, LEONARD. *Elements of Vibration Analysis*. New York: McGraw-Hill, 1975.

NEWMARK, N. M. "Numerical Procedures for Computing Deflections, Moments, and Buckling Loads," *Trans. ASCE*, 108 (1943), 1161–1234.

NEWMARK, NATHAN M., and EMILIO ROSENBLUETH. *Fundamentals of Earthquake Engineering*. Englewood Cliffs, NJ: Prentice-Hall, 1971.

NEWMARK, N. M., and W. J. HALL. *Earthquake Spectra and Design*. Berkeley, CA: Earthquake Engineering Research Institute, 1982.

NEWTON, ISAAC. *Mathematical Principles of Natural Philosophy.* 2nd ed., 1713, trans. Andrew Motte 1729; rev. trans. Florian Cajori. Berkeley, CA: University of California Press, 1960.

PAZ, MARIO. *Structural Dynamics: Theory and Computation.* New York: Van Nostrand Reinhold, 1980.

RAYLEIGH, LORD (JOHN WILLIAM STRUTT). *The Theory of Sound.* (First edition, 1877) New York: Dover, 1945.

RINNE, JOHN E. "Design of Earthquake-Resistant Structures: Towers and Chimneys," in Robert L. Wiegel (ed.), *Earthquake Engineering.* Englewood Cliffs, NJ: Prentice-Hall, 1970.

RUMMAN, WADI S. "Reinforced Concrete Chimneys," in Mark Fintel (ed.), *Handbook of Concrete Engineering.* New York: Van Nostrand Reinhold, 1974.

SALVADORI, MARIO G., and MELVIN L. BARON. *Numerical Methods in Engineering.* Englewood Cliffs, NJ: Prentice-Hall, 1961.

SCARBOROUGH, J. B. *Numerical Mathematical Analysis.* Baltimore: The Johns Hopkins Press, 1958.

STEINBRUGGE, KARL V. "Earthquake Damage and Structural Performance in the United States," in Robert L. Wiegel (ed.), *Earthquake Engineering.* Englewood Cliffs, NJ: Prentice-Hall, 1970.

STEINBRUGGE, K. V., and D. F. MORAN. "An Engineering Study of the Southern California Earthquake of July 21, 1952, and Its Aftershocks," *Bull. Seism. Soc. Amer.,* 44 (April, 1954), 199–462.

THOMSON, WILLIAM TYRRELL. *Theory of Vibration With Applications.* Englewood Cliffs, NJ: Prentice-Hall, 1972.

TIMOSHENKO, S., D. H. YOUNG, and W. WEAVER, JR. *Vibration Problems in Engineering* (4th ed). New York: Wiley, 1974.

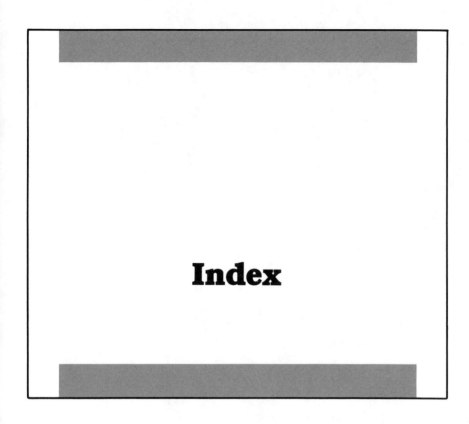

Index